Anwar TON
Ismail BAKLOUTI
Zied DRISS Anouar NJEH

Techno-economic study of the value of integrating photovoltaic systems in vertical buildings

Anwar TON
Ismail BAKLOUTI
Zied DRISS Anouar NJEH

Techno-economic study of the value of integrating photovoltaic systems in vertical buildings

ScienciaScripts

Imprint

Cover image: www.ingimage.com

This book is a translation from the original published under ISBN 978-620-3-43427-9.

Publisher:
Sciencia Scripts
is a trademark of
Dodo Books Indian Ocean Ltd. and OmniScriptum S.R.L publishing group

120 High Road, East Finchley, London, N2 9ED, United Kingdom
Str. Armeneasca 28/1, office 1, Chisinau MD-2012, Republic of Moldova, Europe
Managing Directors: Ieva Konstantinova, Victoria Ursu
info@omniscriptum.com

Printed at: see last page
ISBN: 978-620-8-84383-0

Contents

General introduction

The world's energy needs are constantly increasing. Developing countries in particular will need more and more energy to drive their development. Today, much of the world's energy is produced from fossil fuels. Consumption of these sources leads to greenhouse gas emissions and therefore an increase in pollution. The additional danger is that excessive consumption of natural resources is reducing the reserves of this type of energy in a way that is dangerous for future generations. Renewable energy is energy derived from the sun, wind, the heat of the earth, water or biomass. Among today's renewable energy sources, solar energy is one of the most promising for the future. The Swiss government's 2050 energy strategy is committed to developing photovoltaic solar energy, as well as improving energy efficiency. It is also important to note that a large proportion of primary energy is consumed in the building sector, i.e. in the construction and operation of buildings.

This work is being developed in collaboration with SENCOM (SUSTAINABLE ENERGY COMPANY) and the electromechanical systems laboratory of the Sfax National Engineering School.

The objective of this work is the study and the dimensioning of a photovoltaic installation connected to the network on a large and small scale in the building.

This dissertation comprises four chapters. **The first chapter** presents a bibliographical study of solar energy and existing photovoltaic systems, as well as photovoltaic energy in buildings and their use.

In the **second chapter**, we focus on the technical and economic study of a grid-connected photovoltaic installation designed to supply the Mechanical Engineering Department of the Sfax National Engineering School.

The **third chapter** is devoted to the technical and economic study of a grid-connected photovoltaic installation designed to supply the electromechanical systems laboratory of the Sfax National Engineering School.

In order to put our theoretical knowledge into practice, the fourth chapter focuses on an experimental study of the aerodynamic parameters and energy comfort of the electromechanical systems laboratory at the Sfax National Engineering School.

Finally, we present the conclusion of this work and the prospects suggested by this study.

Our experimental study also focuses on the energy performance of a photovoltaic collector tested during June 2021 in the Sfax region.

Chapter 1

Study Bibliography

1. Introduction

Given the growing demand for electricity, the rising cost of traditional energy sources and the difficulty of connecting Saharan and mountainous regions to the public electricity grid, renewable energy sources - wind power, hydroelectric power, solar thermal and photovoltaic energy, biomass and geothermal energy - offer a solution. What they all have in common is that, when in use, they produce no polluting emissions, thereby helping to combat the greenhouse effect and global warming. These energies are said to be renewable when they are produced by a source that nature constantly renews, unlike energy that depends on sources that run out. There are many different types of energy, but they all come from two main natural sources: the sun and the earth. The sun emits radiation that can be transformed into electricity or heat. It generates zones of uneven temperature and pressure, which are the source of winds. It creates the water cycle and enables plants to grow and generate biomass. The Earth's internal heat can be recovered from its surface. The renewable nature of these energies and their low emissions of waste, pollutants and greenhouse gases are all advantages. But their relatively dispersed energy content is much lower than that of highly concentrated non-renewable energies. In the case of solar and wind power, they can be difficult to store because they are immediately transformed into electrical energy.

Solar energy and photovoltaic energy are generally used for this purpose, especially in the buildings sector, as they are inexhaustible, clean and safe to use. This PV energy is based on the conversion of solar radiation directly into electricity, using solar cells made of light-sensitive semiconductor material.

In this chapter, we study the state of the art in solar energy, focusing mainly on photovoltaic solar energy. More specifically, photovoltaic systems integrated into buildings.

Finally, we end the chapter with a conclusion.

2. Solar energy

2.1 Sun

Solar energy is a clean form of energy that emits no greenhouse gases, and its raw material - the sun - is free, inexhaustible and available everywhere in the world, even though it is more than 150 million kilometres away.

Thanks to its many uses, solar energy tends to be one of the most important energy resources, equivalent to 10,000 times the energy consumption of the whole of humanity. Solar energy is the energy we receive from the sun in the form of electromagnetic radiation, ranging from ultraviolet rays to infrared rays and visible light. This energy heats and illuminates the Earth. It is absorbed by the oceans to create the water cycle, warms the air masses and generates the winds.

The Sun is the only star in the solar system and the closest to Earth. It takes about 8 minutes for its light to reach us. It is the source of all energy, at all times. The Earth receives a flow of solar energy reaching 174 petawatts of power and on an annual basis, the solar energy absorbed by the Earth reaches 3,850,000 exajoules, of which 3,000 EJ are absorbed by the process of photosynthesis. Humanity therefore uses a tiny fraction of this renewable energy in the form of daylight, heat and electricity using solar radiation.

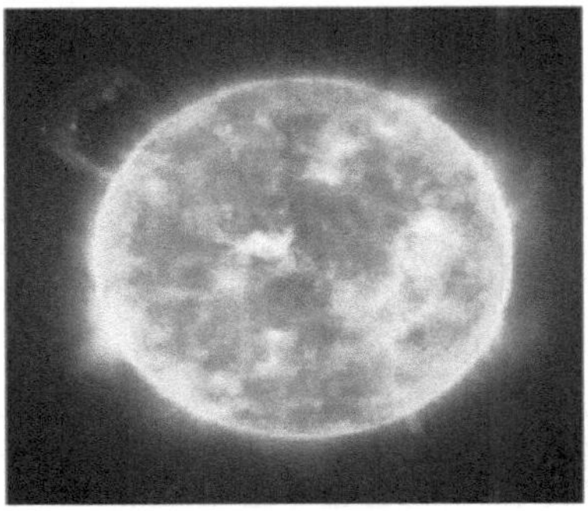

Figure 1.1. The sun.

2.2 Characteristics of solar radiation

At the limit of the Earth's atmosphere, the irradiance received by a surface placed perpendicular to the sun's rays is on average 1,367W/m^2outside the Earth's atmosphere. This is known as the solar constant. Global radiation is of the order of 1,000W/m^2on the Earth's surface. This radiation is the sum of all the radiation received, including the radiation reflected by the ground and objects on its surface. It is measured using a pyrometer or an unshielded "Solari meter".

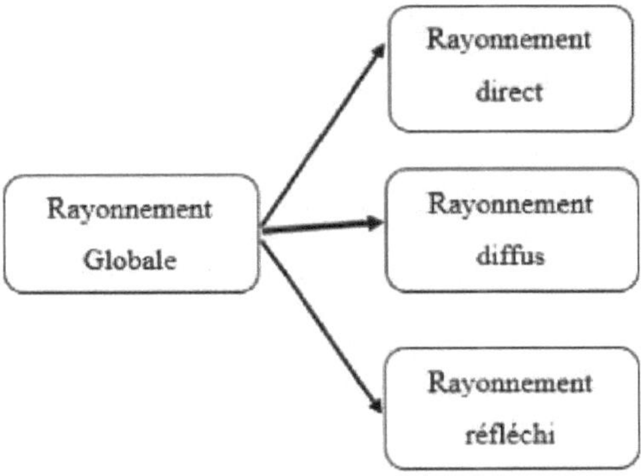

Figure 1.2 Types of solar radiation.

2.2.1 Direct radiation

This is solar radiation formed by parallel rays coming from the sun without being scattered by the atmosphere. Direct radiation is radiation incident on a given plane from a small solid angle centred on the sun's disc. It arrives in a straight line on a clear day.

2.2.2 Diffuse radiation

It is a part of the sun's radiation that has been reflected (scattered) by the molecules and aerosols in the atmosphere. For an observer on the ground, diffuse radiation is a set of radiated directions.

2.2.3 Reflected radiation

The albedo of the ground is the radiation reflected by the ground or by objects on its surface. This albedo can be significant when the ground is particularly reflective (water, snow).

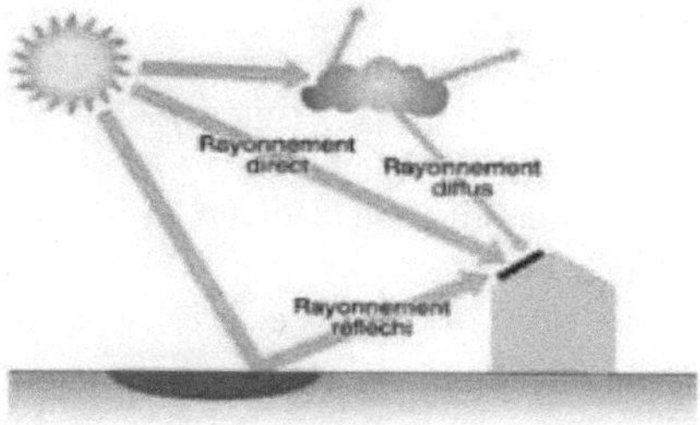

Figure 1.3. Components of solar radiation

2.3 Irradiance or radiation

Illuminance is defined as the power received by a surface. It is expressed in W/m$^{(2)}$ (watts per square metre). The SI (International System of Units) recommends using the symbol E.
Solar irradiance is the energy received by a surface. It is expressed in J/m$^{(2)}$ (Joule per square metre). The ISES (International Solar Energy Society) recommends the symbol H. Other more common units are Wh/m$^{(2)}$ (watt hour per square metre).
On average, Tunisia has more than 3,000 hours of sunshine per year, with some regions enjoying longer hours than others. Most regions in the south of the country have more than 3,200 hours of sunshine a year, with peaks of 3,400 hours a year in the Gulf of Gabès (south-east). On the other hand, the minimum insolation period in the northern regions is between 2,500 and 3,000 equivalent hours of full sun. Solar irradiation varies from 1,800 (kWh)/m^2/year in the north to 2,600 kWh/m^2/year in the south.

2.4 History of solar energy

A brief history of solar energy is as follows:

2.4.1 From the discovery of fire

It was a very long time ago that mankind understood how to use the sun to light a fire. Using mirrors, we can concentrate the power of the sun's rays in one point. The heat then becomes so intense at this point that the fire starts. This is how the ancient Greeks lit the Olympic flame.

2.4.2 Photovoltaics

In the 17th century, a French engineer succeeded in making a pump work using air heated by the Sun. Two hundred years later, a Parisian physicist discovered the photovoltaic effect. It was now possible to convert sunlight into electricity. But it would take another century for this technology to make real progress. It was in the quest to conquer space that scientists invested the most in mastering solar energy.

2.5 Using solar energy

There are two ways of harnessing the Sun as a source of energy:
- Through its heat
- By its radiation

2.5.1 Thermal solar panels

The Sun's heat can be used directly to heat a water tank, dry clothes or cool the walls of a house. This is the principle used by thermal solar panels, which are placed on the roofs or facades of houses. As the water passes through the solar panels, it is heated by the sun and then used on site to heat the building or for domestic hot water.

2.5.2 Thermodynamic power stations

These large installations are equipped with huge parabolic mirrors, spherical or flat, which follow the path of the sun during the day. They concentrate the sun's rays to heat a liquid, such as oil, which remains hot even after sunset. This is a way of storing the energy that will be used to heat water, the steam from which will turn turbines to produce electricity.

2.5.3 Exploiting solar radiation

Sunlight can be used to produce electricity. This is done using photovoltaic panels made up of electronic cells that react to the sun's rays.

In the building sector, solar energy is a relevant response to the context of sustainable development, not only because of its abundance, but also because of the possibilities it offers for natural and efficient lighting, producing useful heat, electricity and even air conditioning.

Of the three ways of converting solar energy: thermal, thermodynamic and photovoltaic, the focus is on photovoltaic solar energy.

3. Photovoltaic solar energy

The term "photovoltaic" is made up of the ancient Greek word "photos" (light, brightness) and the surname of the Italian physicist Alessandro Volta, who invented the electric battery in 1800 and gave his name to the unit for measuring electrical voltage, the volt.

Photovoltaic solar energy is produced by converting sunlight into electricity using photovoltaic cells integrated into panels that can be installed on buildings. The electricity produced can be used on site or fed into the electricity grid.

Figure 1.4. Photovoltaic energy

3.1 History of photovoltaic energy

The history of photovoltaic energy can be broken down as follows:

1839: Discovery of the photovoltaic effect by the French physicist Edmond Becquerel.

1875: Werner Von Siemens presents a paper to the Berlin Academy of Sciences on the photovoltaic effect in semiconductors.

1954: Three American researchers develop a high-efficiency photovoltaic cell at a time when the fledgling space industry is looking for new solutions to power its satellites.

1958: A cell with an efficiency of 9% is developed. The first satellites powered by solar cells are sent into space.

1973: The first house powered by photovoltaic cells is built at the University of Delaware.

1983: The first car powered by photovoltaic energy travels 4,000 km in Australia.

1980s: PV systems used for electrification and water pumping in isolated areas.

Late 1990s: PV systems used to produce electricity fed into the grid in several developed countries (Japan, Germany).

2000s: Favourable policies for PV put in place PV power plants with outputs of tens of MW.

3.2 Photovoltaic cells

3.2.1 Definition

The photovoltaic cell is the smallest component of the photovoltaic installation. It is generally made of silicon, a semi-conductor material extracted from the silica contained in sand. Its role is to convert solar energy into direct current electricity.

3.2.2 Operating principle

A photovoltaic cell is made from two layers of silicon, one P-doped (boron-doped) and the other N-doped (phosphorus-doped), creating a PN junction with a potential barrier. When the cell is exposed to solar radiation, the photons of energy entering the solar cell transmit their energy to the atoms in the junction.

If this energy is high enough, it can cause electrons to move from the valence band to the conduction band of the semiconductor material, creating "electron-hole" pairs. Electrons (negative charges) and holes (positive charges) are kept apart by an electric field that forms a potential barrier. If a charge is placed across the cell, the electrons in the N zone join the holes in the P zone via the external connection, giving rise to a potential difference and an electric current flowing (Figure 1.5).

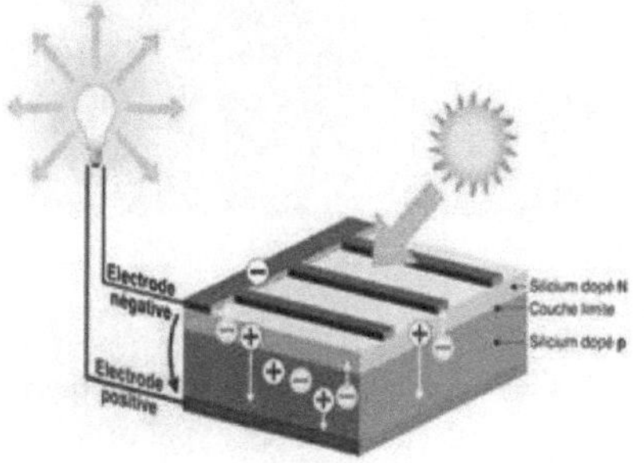

Figure 1.5. Principle of photovoltaic conversion.

3.3 Different types of photovoltaic cells

There are different types of photovoltaic cell. Each type has its own efficiency and cost. There are currently three main types of cell.

3.3.1 Monocrystalline silicon cells

Monocrystalline silicon cells represent the first generation of photovoltaic generators. These cells are generally uniformly blue in colour and offer the highest efficiency of any commercially available solar panel: between 13% and 15%. They are used mainly for space applications.

Figure 1.6. Monocrystalline silicon cells.

3.3.2 Polycrystalline silicon cells

As the silicon cools in an ingot mould, several crystals are formed. The photovoltaic cell is bluish in appearance, but not uniform.

Polycrystalline silicon is used for terrestrial applications, and generally has an efficiency of between 12 and 14%.

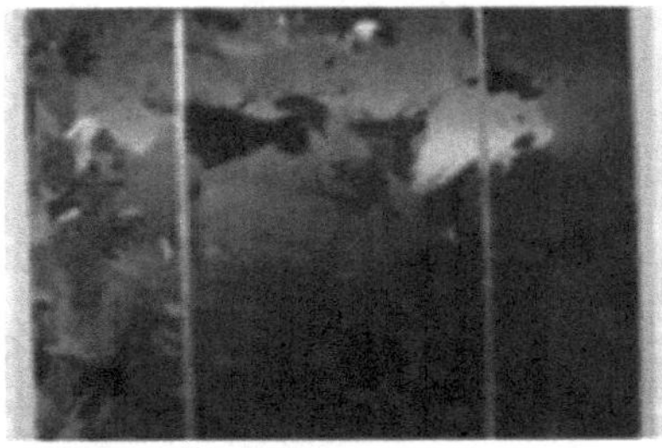

Figure 1.7. polycrystalline silicon cells

3.3.3 Amorphous silicon cells

When silicon is transformed, it produces a gas that is sprayed onto a glass plate, which is why amorphous silicon cells are manufactured. This type of cell is cheaper and easier to manufacture. Their thinness means they can be used to create flexible solar panels. However, these cells have limited efficiency of around 5 to 7%, and are used in calculators and watches.

Figure 1.8. Amorphous silicon cells

Table 1.1 summarises the various characteristics of photovoltaic modules.

Table 1.1 Types of photovoltaic cells

silicon	Monocrystalline	polycrystalline	Amorphous
Module efficiency	14 à 17 %	13 à 16 %	5 à 7 %
Peak power/m²	120 to 150 Wp/m²	110 to 130Wp/m²	60 à 70 Wp/m²

Sensitivity to amount of light	High	High	Lower
Cell colour	Dark blue to black	Blue	Red-brown to blue-violet
Cell appearance	Uniform	mottled	Uniform

3.3.4 Grouping of photovoltaic cells

3.3.4.1 Series array

In this array, the same current flows through the cells and the resulting characteristic of the array is the sum of the voltages at a given current. This array makes it possible to increase the output voltage.

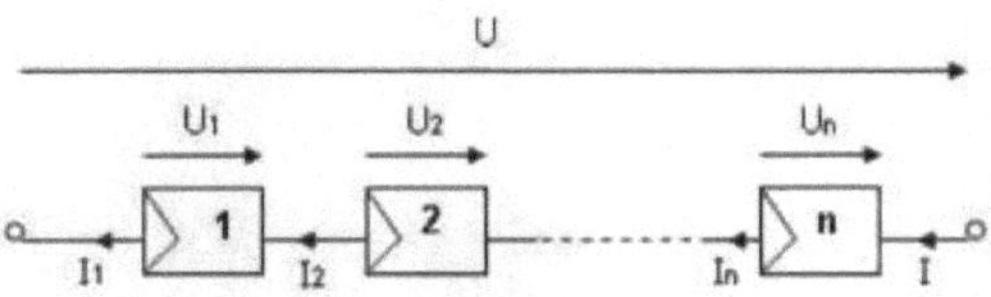

Figure 1.9. Series combination

For an array of n cells connected in series, the general expression for the output voltage Us is :

$$U_s = nU_c \quad (1.1)$$

Where Uc is the voltage supplied by a cell.

3.3.4.2 Parallel grouping

In a group of cells connected in parallel, the cells are subjected to the same voltage. The currents are added together and the resulting characteristic is obtained by adding the currents at a given voltage. This grouping makes it possible to increase the output current.

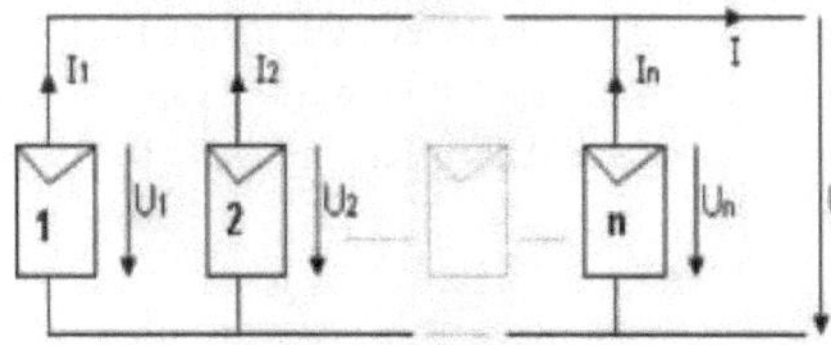

Figure 1.10. Parallel association

For a group of n cells connected in parallel, the general expression for the output current Is is :

$$I_s = n\,I \quad (1.2)$$

Where I is the current supplied by a cell.

All the photovoltaic cells electrically connected together form a photovoltaic module. These cells are found within the photovoltaic panels that make up the photovoltaic systems.

4. Types of photovoltaic systems

A photovoltaic system, or solar photovoltaic system, is an energy system designed to supply solar energy that can be used by photovoltaic systems. It consists of an arrangement of several components, of which the solar panels absorb and convert sunlight directly into electricity. An inverter to convert DC electricity into AC electricity. Mounting, wiring and other electrical accessories are also used.

The PV system is the set of panels connected in series and in parallel to achieve a higher voltage and current. Figure 1.11 shows a cell, a module, a panel and a PV system. Photovoltaic systems are classified into three types: stand-alone, hybrid and grid-connected. The first two are independent of the electricity distribution system. They are often found in remote areas.

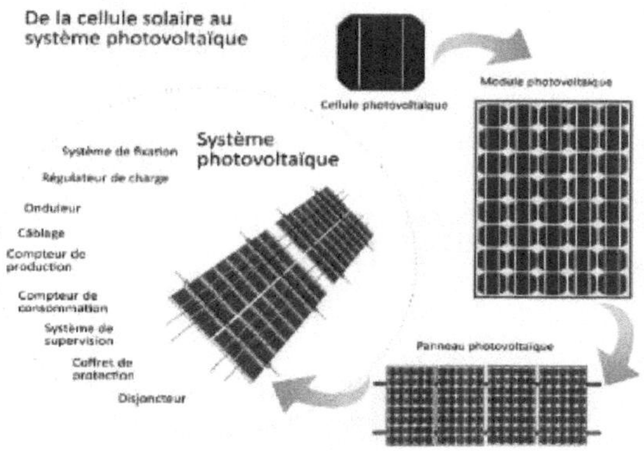

Figure 1.11. From solar cell to photovoltaic system

4.1 Stand-alone photovoltaic system (isolated)

An isolated photovoltaic system supplies the user with electricity without being connected to the electricity grid.

This is often the only way of obtaining electricity when mains power is not available: in isolated locations, on islands or in mountainous areas. This type of system requires the use of batteries to store the electricity and a solar regulator to ensure the durability of the solar batteries. An autonomous photovoltaic installation must be capable of supplying energy even when there is no more sun. Part of the daily production of the photovoltaic modules must therefore be stored.

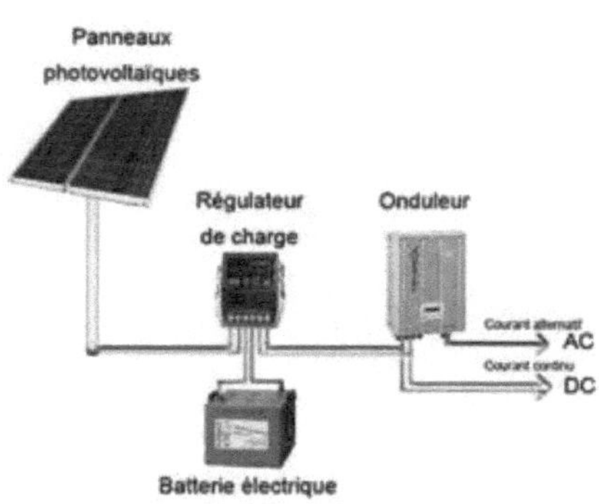

Figure 1.12. Schematic diagram of an autonomous PV installation.

4.2 Hybrid photovoltaic system

Hybrid systems receive part of their energy from one or more additional sources. In practice, PV system modules are often combined with a wind turbine or a fuel generator, or a genset. Such systems usually have energy storage batteries. They are best suited where energy demand is high, where energy needs to be supplied on demand or where the budget is limited.

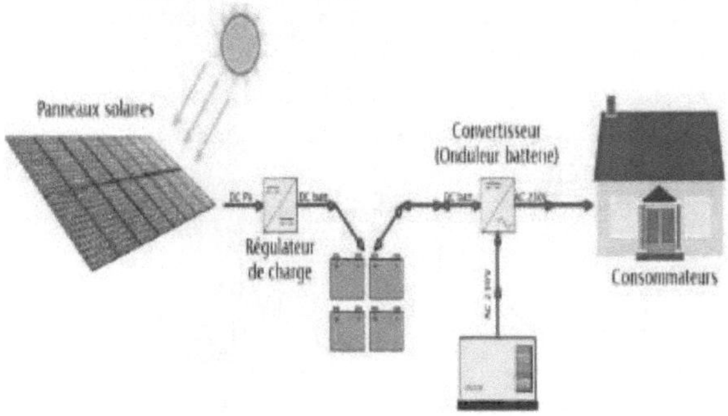

Generator set

Figure 1.13. Schematic diagram of a PV hybrid system

4.3 Grid-connected PV system

Grid-connected installations are systems connected in parallel to the public electricity grid and designed to inject the electrical energy produced into the grid. These installations thus become small "power stations" capable of eliminating or reducing the energy requirements of any public or industrial building or private home.

The system, connected in parallel to the grid, is made up of a specific surface area of photovoltaic panels, suitably interconnected and powered by special devices. The electronic conversion equipment (inverter) adapts the energy produced to single-phase or three-phase grid standards and injects it into the grid. (Figure 1.14).

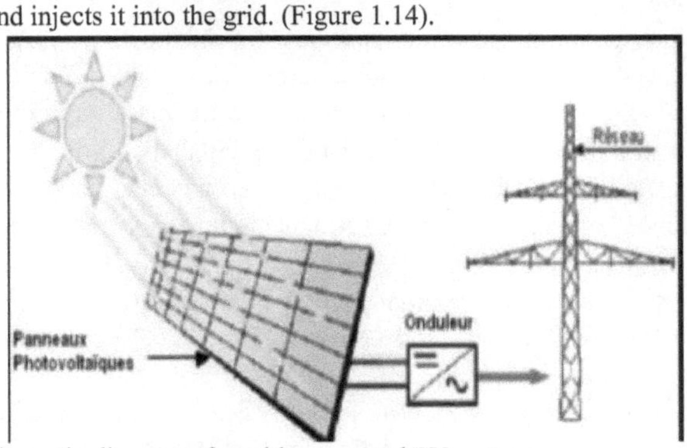

Figure 1.14. Schematic diagram of a grid-connected PV system

4.3.1 Photovoltaic panel or module

The individual cell, the basic unit of a photovoltaic system, produces only a small amount of electrical power, typically 1 to 3 W at a voltage of less than one volt (0.6 V). To produce more power, the cells are assembled to form a PV module or panel.

4.3.1.1 Solar resources

To produce the maximum amount of electricity, a photovoltaic panel must be positioned perpendicular to the sun's rays. This is generally impossible to achieve, as the sun's position varies constantly depending on the time of day and the time of year.

4.3.1.2 Tilt

Inclination is the angle made with the horizontal plane. The optimum angle in Sfax is 30° so that solar radiation is perpendicular to the panels.

The tilt of the panels is important to optimise the irradiation of the direct and diffuse radiation received by the collectors. The more the surface of the solar panel is perpendicular to the sun's incident rays, the greater the efficiency.

Table (1.2) shows the effect of tilt on PV module output.

Table 1.2: Correction factor for a given tilt angle

INCLINAISON / ORIENTATION	0°	30°	60°	90°
EST	93%	90%	78%	55%
SUD - EST	93%	96%	88%	66%
SUD	93%	100%	91%	68%
SUD - OUEST	93%	96%	88%	66%
OUEST	93%	90%	78%	55%

In general, the tilt angle of photovoltaic panels is between 15 and 35°. In Tunisia, the best tilt angle for capturing the maximum possible solar irradiation throughout the year is 30°.

4.3.1.3 Orientation

Orientation is the angle between the solar panel and the due South axis. At 0°, the angle of orientation corresponds to a panel facing south, at 90° east or west and at 180° north. The orientation of the solar panels is important to capture as much sunlight as possible and produce as much energy as possible. This way, the installation will pay for itself more quickly. Solar panels always face south to capture the maximum amount of sunlight and produce the maximum amount of electricity. Although ideal, this orientation is by no means compulsory.

In Tunisia, and more generally in the northern region, facing due south is the best possible orientation for a photovoltaic module. This is how it will produce the most electricity.

4.3.1.4 Shading

Photovoltaic modules are very sensitive to shadows. Shading means that a mask prevents photons from reaching the photovoltaic cells. Total or partial shading conditions have a significant impact on energy production capacity and can lead to losses. The cells of a solar panel are generally connected in series to obtain more power, and therefore suitable electricity production. However, when shading occurs, this structure has certain limitations.

The polarity at the terminals of the shaded cell can be reversed (the voltage changes sign): the cell no longer behaves as a generator but as a receiver. The cell then heats up, which can cause irreparable damage. This phenomenon of the cell heating up is called a hot spot.

Shade is a formidable enemy for a module, resulting in a significant loss of power and heating of the cells.

4.3.1.5 Electrical characteristics of a photovoltaic panel

A photovoltaic panel is made up of a set of photovoltaic cells. In the presence of sunlight, each of these cells produces a current I (A) and a voltage U(V).the product of these two quantities gives a power P, expressed in watts (W).

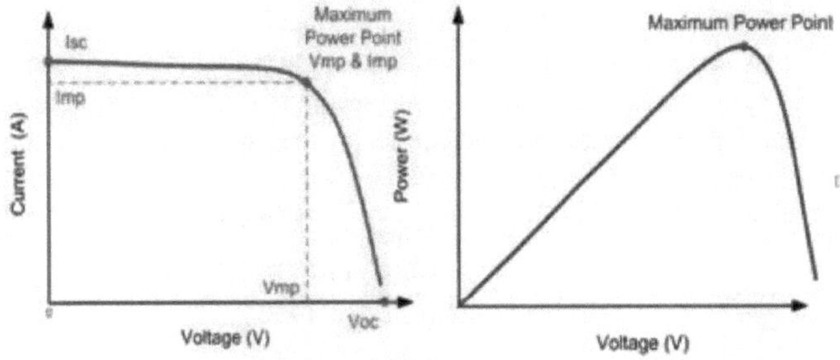

(a) Current-voltage characteristic (b) Power-voltage characteristic

Figure 1.15. Current-voltage characteristic and power of a PV panel

An example of the electrical characteristics of a photovoltaic panel is shown in Figure 1.15.

For the current-voltage curve ;

The open circuit voltage V_{oc} indicates the voltage present across the panel when it is not connected.

The short-circuit current I_{sc} (Short Circuit) indicates the maximum current that the panel can deliver.

The power curve is obtained by multiplying the current and voltage of the current-voltage curve.

The Maximum Power Point (MPP) on this curve corresponds to the product of two specific values: the maximum power current (I_{mpp}) and the maximum power voltage (V_{mpp}).

The power supplied to the external circuit by a photovoltaic cell under illumination depends on the load resistance. This power is maximum for an operating point Pm ($I_m = I_{max}$, $V_m = V_{max}$) of the current-voltage curve.

The form factor FF is defined by the following relationship:

$$FF = P_m \, V_{CO} \, I_{CC} \qquad (1.3)$$

With :

$$P_m = V_m \, I_m \qquad (1.4)$$

The conversion efficiency η of the photovoltaic cell is expressed as a percentage % and refers to the power conversion efficiency. It is defined as the ratio between the maximum power delivered by the solar cell and the incident power Pin.

The yield is thus defined by :

$$\eta = P_m / P_{in} \qquad (1.5)$$

or :

$$\eta = FF / V_{CO} \, P_{in} \qquad (1.6)$$

There are several parameters that influence the efficiency of photovoltaic modules.

4.3.16 Influence of irradiance

The PN junction reacts differently depending on the energy it receives. The more energy it receives, the more it releases. For a photovoltaic installation, a 50% variation in illuminance or load leads to a deterioration in the power supplied by the PV generator of the order of 50%, and the PV generator no longer operates under optimum conditions.

Figure 1.16 shows the variation of the current-voltage and power-voltage characteristics as a function of illuminance. It can be seen that brightness has a considerable influence on module performance. The short-circuit current ICC varies proportionally with the illuminance E and the open-circuit voltage VCO varies very little with the illuminance. The influences of these parameters therefore result in variations in the useful characteristic of the photovoltaic generator with the lighting conditions. Thus, the greater the cloud cover, the lower the intensity of the current generated.

N°	Designation
1	$E=1000W/m^2$
2	$E=800W/m^2$
4	$E= 600W/m^2$
5	$E=400W/m^2$
6	$E=200W/m^2$

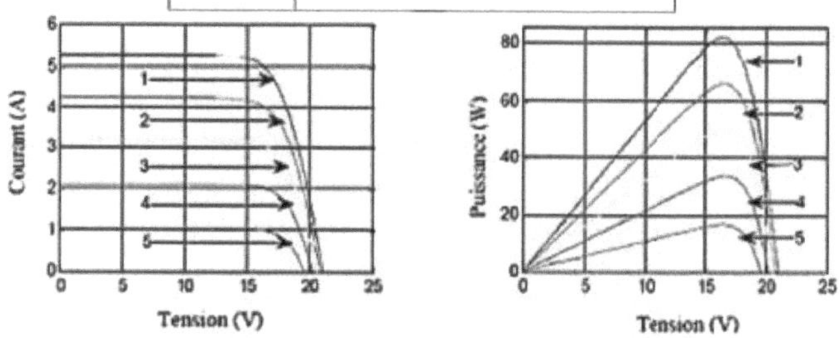

(a) Current-voltage (b) Power-voltage

Figure 1.16. Influence of illuminance on the current and power of a PV module.

4.3.1.7 Influence of temperature

Temperature is a very important parameter in the behaviour of solar cells. It also has an influence on the characteristics of a PV generator. Figure (1.17) shows the effect of temperature on the current-voltage and power-voltage characteristics. Temperature significantly affects the operation of the PV module. In addition, the temperature coefficients differ depending on the materials used to manufacture the module. This influence must therefore be taken into account when sizing the photovoltaic generator. Under these conditions, it can be seen that the open-circuit voltage VCO decreases with temperature, while the short-circuit current ICC increases only slightly with temperature. The influence of temperature on Ice can be neglected in most cases.

N°	Désignation
1	55°C
2	44°C
3	40°C
4	32°C
9	25°C

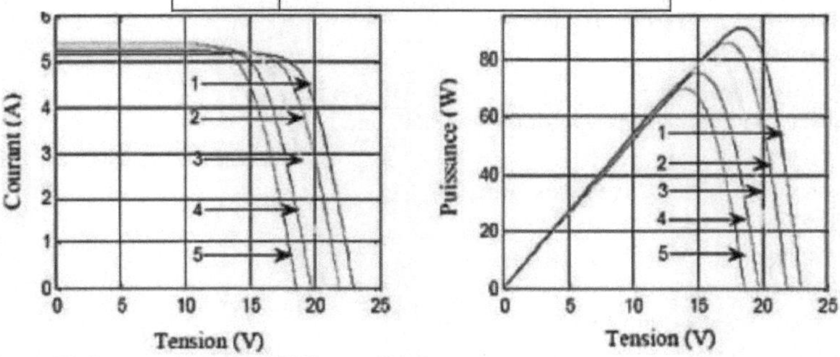

(a) Current-Voltage (b) Power-Voltage

Figure 1.17. Influence of temperature on the (I-V) and (P-V) characteristics

4.3.2 Photovoltaic inverters

An inverter performs several essential functions in a grid-connected PV installation:

- Converting DC current and voltage into grid-compatible AC current and voltage.
- Finding the maximum power point of the photovoltaic field.
- Decoupling protection.
- Checking the insulation of the DC part of the photovoltaic installation.

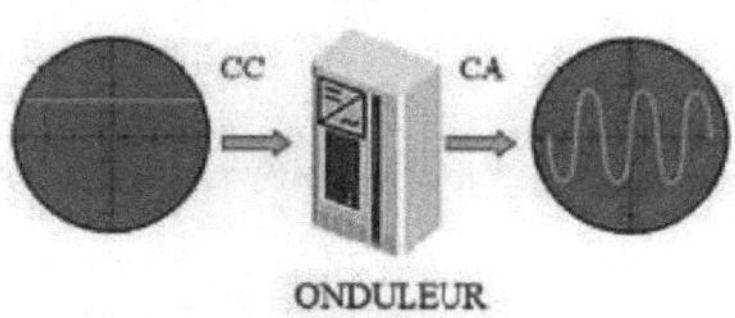

Figure 1.18. Photovoltaic inverter

4.3.3 Protective devices

4.3.3.1 Junction box

A junction box is :

- "A Susceptible" to contain fuses on each of the photovoltaic strings;
- Located as close as possible to the photovoltaic field;
- Allows photovoltaic strings to be connected in parallel;
- Has breaking and sectioning capacity;
- Protection against overvoltages if required.

4.3.3.2 DC/AC protection box

There are two types of protection box, one for the DC side and the other for the AC side.

A DC electrical cabinet is :

- Located upstream and as close as possible to the inverters
- Contains a switch-disconnector and a DC lightning arrester
- Has breaking and sectioning capacity Protection against overvoltages.

An AC electrical cabinet is :

- Located downstream and as close as possible to the inverters
- Contains an AC surge arrester and a differential circuit breaker
- Has breaking and sectioning capacity
- 30 mA phase-to-neutral differential circuit breakers
- Protection against overcurrents.

4.3.3.3 Fuses

The purpose of the fuse is :

- protect electrical circuits against short-circuits and
overloads
- Fusing if the current exceeds its rating In
- Operating voltage: U or Ue
- Breaking capacity: PC > ISC

4.3.3.4 Circuit-breakers

Circuit-breakers are designed to :

- To make, carry and break currents under normal circuit conditions.
- To switch on, withstand for a specified time and interrupt currents under specified abnormal conditions: short circuit or overload.

The characteristics of circuit-breakers are :

- Ue: Rated operating voltage is the maximum operating voltage
- In: Rated current: maximum value of continuous current
- ICU: Ultimate breaking capacity or rated breaking capacity I_{cn}
- The maximum theoretical short-circuit current which the circuit-breaker is capable of interrupting without risk of destruction. Value higher than that which can be reached on the line.
- Limiting capacity: the ability to let through only part of the current in the event of a short-circuit.
- Number of poles interrupted and number of poles protected (4P, 3P): depends on the network and the load.
- Model (modular, compact, etc.): mainly dictated by In.

4.3.3.5 Disconnector

The disconnector is a mechanical connection device capable of :

- Opening and closing a circuit when the current is zero or practically zero (no breaking capacity).
- Isolating the part of the installation downstream of the disconnector.
- Safety device during maintenance work.
- The disconnector characteristics are :
- Maximum current supported by the power poles:
- Maximum insulation voltage between the power poles ;
- Number of power poles and pre-break contact;
- Possibly with or without fuse melting detection;
- Mostly installed at the head of an electrical installation;
- Can be operated from outside the electrical cabinet using a handle. Unlike the switch-

disconnector, the fuse-holder disconnector has no breaking capacity: it cannot disconnect an electrical circuit under load.

4.3.3.6 Lightning arresters

Surge arresters are designed to

- dissipate the overcurrent (without it passing through the equipment to be protected)
- To limit the voltage (so that the equipment to be protected does not break down). The types of surge protector are :

Type 1 surge protectors: with a very high flow capacity. They are designed to protect the heads of buildings fitted with lightning conductors.

Type 2 surge protectors: with a high flow capacity. They are used for head protection in the absence of a lightning conductor.

Type 3 surge protectors: these are used exclusively for the fine protection of receivers and are installed behind a type 1 or type 2 surge protector.

4.3.4 Electrical wiring

Wiring is used to make the electrical connections between the various components. There are three types of wiring:

- Wiring on the DC side (solar cables): developed to meet the specific requirements of photovoltaic solar energy.
- Wiring on the AC side: complies with the rules applicable to standard electrical installations
- Wiring for earthing and equipotential bonding of all components.

4.3.5 Meters

A bi-directional meter will be used to record the amount of energy supplied to and extracted from the grid.

5. Use of photovoltaic systems

The purpose of photovoltaic systems is to produce electricity, which can be used in a number of applications. A first way of classifying the applications of photovoltaic energy is to distinguish between applications connected to the electricity grid and isolated installations.

5.1 Applications in isolated systems

In isolated systems, i.e. systems that are not connected to the electricity grid. The range of applications for photovoltaic energy is almost infinitesimal. Nevertheless, here are a few examples:

- Lighting systems.
- Signalling using photovoltaic energy.
- Photovoltaic systems linked to telecommunications.
- PV energy on farms and livestock farms
- Photovoltaic systems for cottages and houses

5.1.1 Photovoltaic lighting systems

Street lighting using photovoltaic systems is presented as one of the most economical solutions for lighting town entrances and level crossings. This type of photovoltaic application can also be found on advertising hoardings, bus stops, street lamps and tunnel lighting, among other examples.

5.1.2 Signage using photovoltaic energy

Photovoltaic energy is used to automate lighthouses and buoys for maritime use. Another major application is the signalling of roundabouts, curves, road signs and obstacles using high-brightness LEDs. The low power consumption of LEDs means that a small photovoltaic installation can be used in these systems.

5.1.3 Photovoltaic systems linked to telecommunications

In the telecommunications sector, photovoltaic energy is used for mobile telephony, radio and television repeaters, roadside radar, radiotelephony and military or forestry surveillance posts. Rural telephony by satellite is an example of the use of solar energy.

5.1.4 PV energy on farms and livestock farms

In livestock farming applications, photovoltaic energy is used to power pumps to supply water to cattle troughs. In agricultural applications, photovoltaic systems are used effectively around the world to pump water for livestock, plants and humans. This is because the need for water is greater on hot, sunny days.

5.1.5 Photovoltaic systems for cottages and houses

Photovoltaic systems are an economical option for cottages and houses in remote areas. In most of these areas there is no connection to the electricity grid and in many cases expensive fossil fuels are used to generate electricity. These photovoltaic systems are also used in isolated buildings. This use is very useful because the investment required to place solar panels on the roof of a farm, a chalet in the mountains or a hotel in a remote area, is much lower than it would cost to connect to the electricity grid (1).

5.2 Applications of grid-connected systems

Grid-connected photovoltaic installations are one of the most recent and innovative applications of photovoltaic systems. Here are a few examples of applications for grid-connected photovoltaic systems (2)

- **PV panel installations on roofs**

In homes connected to the electricity grid, the surface area of the roof makes it possible to install easy-to-install modular systems.

- **Production facilities**

Photovoltaic installations are industrial applications that can be installed in rural areas not used for other purposes, or superimposed on large urban areas.

- **Street and road lighting**

Independent systems to power small appliances or for street and road lighting. In transport Some electric vehicles have integrated solar systems to drive an electric motor and obtain motive power.

6. Integrating photovoltaic systems into buildings

It involves replacing traditional architectural elements with new architectural elements that include photovoltaic elements (generally photovoltaic solar panels), and which are therefore power generators. Sometimes they can be combined with architectural designs to take advantage of passive solar energy, or with solar thermal systems. Given the current energy crisis, solar photovoltaic systems are becoming increasingly popular. in buildings, especially vertical buildings, has become extremely important.

The building sector has embarked on a profound transformation to meet new environmental and economic requirements, resource savings, energy efficiency and smart grids. Today's major challenges revolve around one major objective: the construction and management of buildings using renewable energy sources, primarily solar power.

6.1 Types of technology for capturing solar energy

Solar energy is currently collected using three main technologies:

- Photovoltaic
- Concentrated thermodynamic solar power plants.
- Solar heating and cooling systems

6.2 Reasons for using solar technologies

Solar energy is used in a wide range of devices, from calculators to cooking and from agricultural technology to space technology. But in the space of just a few years, its use in buildings has taken off. There are several reasons why buildings are increasingly opting for solar sources:

- Increasing energy consumption
- Growing environmental awareness
- Climate change
- Attraction to cleaner, renewable energy sources
- High social value and increased property value

6.3 Why buildings are turning to solar energy

Solar energy is a renewable, clean and freely available source of energy. Per capita energy consumption is increasing at an alarming rate. Energy prices are also reaching record levels. Buildings are now being constructed with energy efficiency in mind. The aim is to minimise energy consumption and use alternative solar energy, which has many advantages.

There are four aspects of energy efficiency in a building where technology can play a key role:

- Before construction: designing a passive building with zero energy consumption.
- During construction: using energy-efficient building materials.
- Low-energy equipment: to reduce operational energy requirements to a minimum.
- Integration with other applications: the integration of renewable energy technologies for various applications.

6.4 Advantages of using solar energy

There are a number of advantages to using solar energy in buildings, including

- Unlimited resources : Solar energy is an inexhaustible renewable energy.
- Low impact on the environment: Depending on the size of the system installed, whether rooftop photovoltaic panels or large-scale power plants, solar technologies can generate less pollution for the environment.
- Energy independence: it makes buildings energy independent and reduces pressure on the use of natural energy sources.
- Multi-use: it can be used in many different ways and for many different applications.
- Expandable: Photovoltaic systems can be easily expanded because they are modular.
- Portable: Easy to transport.
- No costs after installation: Once the infrastructure has been installed, there will be no further costs.

6.5 Ways of reducing energy consumption

There are three ways of reducing the energy consumption of buildings:

- Solar-powered buildings improve the efficiency of the heating and cooling system by 30% by installing a new system. The operating cost of a high-efficiency solar water heater can be reduced by 90% in energy-efficient solar-powered buildings.
- Using windows with low-emissivity glazing can reduce the space's cooling requirements by around 40%. A light-coloured roof reduces the temperature of a roof because it absorbs less than 50% of the sun's energy.

- The use of low-energy lights and electronic devices can reduce a solar building's energy consumption by 20-30%. With the continuing emergence of technological innovations in the field of construction, it is becoming easier and easier to improve the energy efficiency of buildings.

6.6 Integrating photovoltaic energy

Buildings occupy a central place in our daily lives: we spend a large part of our days in them, whether at home, at work or in our free time. The building environment, which takes many different forms - homes, workplaces, schools, hospitals, libraries and other public buildings - is the biggest consumer of energy and one of the main emitters of carbon dioxide.

Collectively, buildings account for 40% of energy consumption and 36% of greenhouse gas emissions, which come mainly from construction, renovation and demolition activities, as well as from the use of buildings.

Improving the energy efficiency of buildings therefore has a key role to play in achieving Europe's ambitious goal of carbon neutrality by 2050. The building is becoming an energy producer in industrialised countries, consuming a significant proportion of the energy used, essentially photovoltaic energy, which is part of an architectural and functional integration, it is advisable to anticipate the integration of the system from the design of the building and/or the photovoltaic PV installation.

6.6.1 Definition of PVIB

Building Integrated Photovoltaics (BIPV) refers to photovoltaic modules that replace conventional building materials while producing energy. Building integrated photovoltaics (BIPV) are photovoltaic installations that replace the traditional construction elements of houses and buildings. BIPV is a technology that consists of replacing certain structural elements of the building or its surroundings with other elements that incorporate photovoltaics. They are products or systems for producing solar energy that are seamlessly integrated into the building envelope, making up part of the building such as a roof, facade or windows. BIPV systems are integral components of building envelopes and convert solar energy into electricity while fulfilling building envelope functions such as:

- Weather protection.
- Thermal insulation
- Noise protection
- Daytime lighting
- Safety(4)

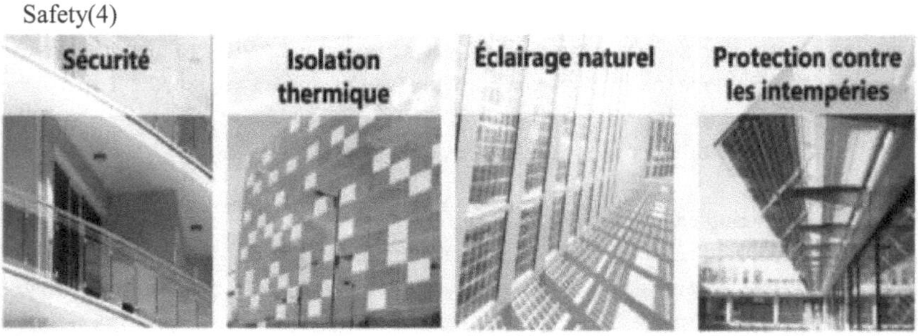

Figure 1.19. Functions of the building envelope.

6.6.2 History

The integration of photovoltaics into buildings began in the 1980s in the United States. These systems, installed on buildings connected to the grid, are not superimposed but integrated aesthetically, because photovoltaics are coming to town and are no longer confined to isolated sites. This concept first appeared around fifteen years ago in Joule II's European Solar House programme·In the 1990s, the production of modules and systems that could be integrated into buildings began as demand increased, but it was not until the 2000s that this sector really took off thanks to the integration premiums on the feed-in tariff. While the primary objective of the conventional sector is to reduce the W/€ ratio of solar panels, the PVIB building-integrated sector is developing the aesthetics and coherence of photovoltaic installations in their environment.

6.6.3 The PVIB market in Tunisia

In Tunisia, electricity consumption in the buildings sector accounts for 27% of final energy consumption. This percentage is set to rise over the next few years. The building sector represents a high potential for energy savings, which could reach 56% by 2030. According to forecasters, the installed capacity of PVIB should increase from 343.1 MWp in 2012 to 1,152.3 MWp in 2019, representing an average annual growth rate of 18.7%, and another estimates that the building market should grow between 2015 and 2022 from 3 billion$ to more than 26 billion $.

The building-integrated photovoltaic system installed in Rades by Aurasol in collaboration with the Tunis Chamber of Commerce and Industry, a partner in the FOSTEr in MED project, a building-integrated photovoltaic solar system has been installed in Tunis. The system has a power output of 20 kWp (watt-peak) and is made up of 80 panels that fit in perfectly with the architecture of the existing building.

The system is based on renewable energy sources, namely solar power.

In the other partner countries of the Programme, the achievements remain similar and keep the same spirit of the project. In the case of Lebanon, the project's official website states that the Industrial Research Institute (IRI) in Lebanon has presented its pilot solar project in Beirut, installed on the IRI building on the Lebanese University campus. The project exceeds the planned 17 kWp, with a 30 kWp system of photovoltaic panels as solar protection on the facades and as cladding on the roof of the building, with an annual production of 36 MWh/year. At all times, a monitoring system records all the energy efficiency parameters, such as power, panel temperature, energy production and daily CO_2 emissions.

6.7 Different types of building

The term "building-integrated photovoltaics" is sometimes used to refer to photovoltaic systems that are integrated into the building after construction. These PVIB systems have low running costs and can be installed in any type of building, including commercial, industrial, institutional, agricultural and residential. One example is integrating photovoltaics into vertical buildings (apartment blocks).

6.8 Different types of integration of PVIB systems

BIPV systems can be installed during the construction phase of a building or as part of renovations to an existing building and replacement of envelope components. The building environment allows PVIB systems to be integrated in a number of ways.

Photovoltaic panels can be used in any part of the building envelope. Although the roof is currently the preferred area for integrating PV modules because of the more advantageous solar capture and ease of placement, facades and windows often offer greater potential. This new generation of photovoltaic solar panels completely changes the architectural approach, with a "cladding" that plays with shapes and colours, and makes it possible to install very discreet and functional panels that produce electricity. There are four main types of PVIB product: roofs, facades, glazing, and transparent and translucent photovoltaics.

6.8.1 Roofs

Photovoltaic solar panels produce electricity and are ideally installed on the roof of a building, which offers the necessary inclination for the best yield: **Inclination**: the panel must be inclined at a precise angle (around 30° depending on the type), for optimum yield. In general, the roof of a building is ideally tilted for installing a photovoltaic or thermal solar panel.

Safety: when installed on a roof, solar panels are less accessible and therefore better protected from potential accidents. On the other hand, they are a little more complicated to maintain, as a ladder is required.

Space: solar panels are bulky (at least 2 $m^{2)}$. Ideally, roofs have a large surface area available for installing photovoltaic panels.

For buildings, we describe two types of rooftop photovoltaic integration as: inclined, roof-integrated systems and non-inclined (vertical) systems.

Pitched, roof-integrated systems may be inaccessible or accessible from inside the building.

Vertical **systems are** accessible from inside the building.

There are two main ways of integrating rooftop photovoltaics into PVIB buildings.

6.8.1.1 Flat roofs

The most widely installed to date is an amorphous thin-film solar cell integrated into a flexible polymer module that has been attached to the roofing membrane using an adhesive sheet between the backsheet of the solar module and the roofing membrane. CIGS technology is now capable of delivering a cell efficiency of 17%.

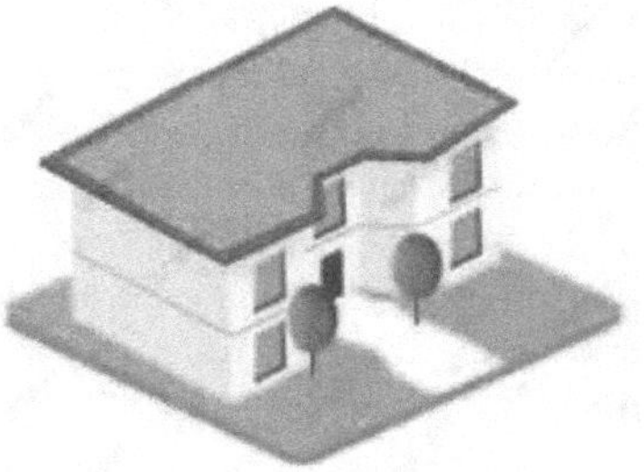

Figure 1.20. Buildings with flat roofs.

6.8.1.2 Pitched roofs

Solar tiles are (ceramic) tiles with integrated solar modules. The ceramic solar tile was developed and patented by a Dutch company in 2012.

Solar shingles are multi-tile modules designed to look and act like ordinary shingles, while incorporating a flexible thin-film cell.

It extends the normal life of the roof by protecting the insulation and membranes from ultraviolet radiation and water degradation.

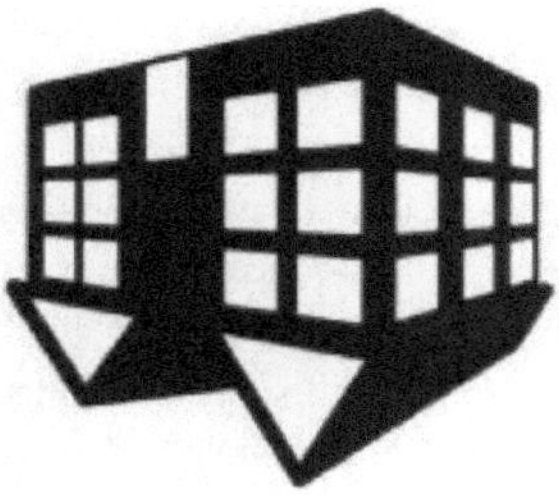

Figure 1.21. Buildings with pitched roofs.

6.8.2 Facades

Facades can be installed on existing buildings, giving old buildings a whole new look. These modules are mounted on the façade of the building, on top of the existing structure, which can increase the attractiveness of the building and its resale value.

Photovoltaic panels replace the façade cladding on traditional façades or curtain walls. Usually, a cavity is left behind the panels to allow ventilation and water run-off without affecting the façade insulation.

Figure 1.22. BIPV in facades.

6.8.3 Glazing

Photovoltaic panels act as glazing. The density of the cells allows a greater or lesser degree of transparency, which can also have a solar protection function.

Photovoltaic windows are semi-transparent modules that can be used to replace a number of architectural elements commonly made from glass or similar materials, such as windows and skylights. As well as producing electrical energy, they can generate additional energy savings thanks to superior thermal insulation properties and solar radiation control.

6.8.4 Transparent and translucent photovoltaics

Transparent solar panels use a coating of tin oxide on the inside surface of the glass to draw current out of the cell. The cell contains titanium oxide coated with a photoelectric dye.

Most conventional solar cells use visible and infrared light to generate electricity. In contrast, the innovative new solar cell also uses ultraviolet radiation. Used to replace conventional glass or placed on top of glass, the mounting area can be large, leading to potential uses combining the functions of energy production, lighting and temperature control.

6.9 Advantages of BIPV systems

The advantages of BIPV systems are numerous. Building-integrated photovoltaic systems generate clean electricity on site without requiring additional land space, but they can also have an effect on a building's energy consumption by using daylight and reducing cooling loads.

BIPV systems can therefore contribute to net zero energy buildings. By transforming roofs and façades into energy producers, BIPV systems are the only building materials that offer a return on investment. What's more, the diverse uses of BIPV systems offer architects and building designers plenty of scope for embellishing buildings. Building owners also benefit from lower electricity bills and a "green" and "innovative" image.

6.9.1 Comfort

Conformity is the quality of something that is the exact representation of a living being, a thing, an activity. Or of a moral or social rule. The word "conform" expresses more than a resemblance; it expresses a perfect identity. In geometry, we can speak of superimposable figures. Comfort is a state of general well-being. It is measured a contrario by the dissatisfaction rate of occupants. Independently of the conditions specific to the individual, it is recognised that the following parameters play a part in comfort:

- Thermal conditions
- Air quality
- Acoustics
- Optics

A comfortable building provides its occupants with a pleasant indoor climate that is less dependent on external conditions, particularly weather and acoustic conditions. Optimum energy management, although very important from the point of view of its impact on the environment, is a second priority, once comfort conditions have been met.

6.9.2 The environment

The conversion of sunlight into electrical energy is renewable and sustainable. Today's photovoltaic panels pay for themselves in less than two years. Energy is generated without any moving parts, which means no maintenance and no noise. This is also true for photovoltaic installations on supports placed later on the roofs of buildings. If photovoltaics are to make a significant contribution to the supply of electricity, large surfaces will be required. For this reason, a large proportion of the exterior surface must be dedicated to photovoltaics. According to estimates by the International Energy Agency, 30% of electricity needs in Germany and 33% in the Netherlands can be met by photovoltaics on the exterior surface of buildings (facade, roof). If new university buildings are to be environmentally pioneering, they need to provide as much surface area as possible for photovoltaics. This undoubtedly includes roofs, but also south-facing facades, which still offer 65% of the energy potential of optimally oriented surfaces. For this reason, it is essential that the façades are well insulated.

6.9.3 Aesthetics

The BIPV system should be integrated into the design of the building from the outset. In this way, attractive aesthetic solutions can be found. If the photovoltaic panels are installed on the roof after the building has been completed, they may be the subject of complaints. For this reason, the tender must include the integration of photovoltaics (PVIB) from the outset.

6.9.4 Technical specifications

If the photovoltaic installation is planned from the outset, technical integration will be simpler. Space for rectifiers, lightning conductors and ducts can be provided in advance. The façade of the building must not be damaged at a later date.

6.9.5 Cost savings

Because the photovoltaic system integrated into the building replaces some of its components, such as part of the roof or facade, the integrated solution is less expensive. Functions normally carried out by the building's facade include protection against the weather and insulation against noise and heat.

6.9.6 cost

It is extremely difficult to estimate the cost of BIPV, mainly because of

- the constantly changing cost of PV modules
- the diversity of technologies used, which is constantly evolving
- the diversity of applications fulfilled by PVIB systems
- the diversity of the components it replaces and the support materials used
- the diversity of characteristics (shape, colour, transparency, etc.).

Some applications cost more than conventional PV, while others cost less, making them more attractive.

7. Conclusion

In this chapter, we are interested in the presentation of the state of the art referring to polar energy and the different applications that are related to it. A description of the principle of photovoltaic systems is also given. In particular, we are interested in the different applications being considered for integrating photovoltaics into buildings. Solar-powered buildings are particularly effective at conserving energy in a variety of ways. There is no doubt that buildings powered by solar energy, with the right technologies, are more energy efficient. Our aim, therefore, is to integrate photovoltaic systems into buildings that provide a certain level of energy comfort.

Chapter 2

Calculation and sizing of a large-scale photovoltaic installation

1. Introduction

In this chapter, our work consists of sizing a photovoltaic installation connected to the medium-voltage network and integrated into vertical buildings, using the mechanical engineering department of the Ecole Nationale d'Ingénieurs de Sfax (ENIS) as a model. As a first step, we calculate the energy requirements and peak power for this department. Next, we select the necessary equipment, protection components and cable cross-sections. Finally, we validate our study with a simulation using PVSYST software.

2. State of play

The study of a photovoltaic installation is carried out by choosing the Mechanical Engineering Department of the Sfax National Engineering School (ENIS) as the subject of our study.

2.1 Presentation of the site

The study of a photovoltaic installation is carried out in the Mechanical Engineering Department at the National Engineering School of Sfax at Cité AL Habib.

The Ecole Nationale d'Ingénieurs de Sfax (ENIS) is a Tunisian engineering school attached to the University of Sfax. It was created in 1983 by a law transforming the Sfax Faculty of Science and Technology (created in 1975), and plays a major role in the development of a regional scientific centre.

(a) Location map

(b) Front facade **(c)** Rear facade

(c) Left side **(d)** Right side

Figure 2.1. Selected site . Description of the Mechanical Engineering Department

The Mechanical Engineering Department at the National Engineering School in Sfax is located on the Soukra road, km 4, next to the Faculty of Science. The various characteristics of the department are summarised in table 2.1 :

Table 2.1. Characteristics of the department

Location	Soukra road Sfax
University	Sfax National Engineering School
Altitude	34.725614462842195m
Longitude	10.717552856318858m

It has three floors and contains various rooms, laboratories, offices and toilet blocks, as shown in table (2.2).

Table 2.2. Components of the Mechanical Engineering Department

Level	**Components**
Ground floor	1 Administration room
	4 laboratories
	1 computer room
	2 teaching offices
	2 toilets
First floor	4 computer rooms
	4 teaching rooms
	2 teaching offices
	4 toilets
Second floor	5 teaching rooms
	1 computer room
	2 teaching offices
	4 toilets

3. Calculating electrical requirements

The first step was to determine the electricity requirements and consumption profile of the Mechanical Engineering department. To do this, we determined all the consumption points, the quantity of electricity consumed and their operating times.

In a building, the main electricity-consuming items are lighting and electrical equipment and electrical auxiliaries.

A simple calculation allows us to easily determine the energy consumption of appliances in kWh. The word kWh owes its name to James Watt, and is a unit of energy written kilowatt-hour in the International System of Units (SI), so kilowatt and watt are units of power, while kWh is a unit of energy. The kilowatt is used in the electrical industry to indicate the power of a machine and the watt indicates the power of an appliance such as a small or large household appliance. In contrast, the kWh indicates the amount of energy consumed by an appliance with a power of 1 kW over a period of one hour.

3.1 Electrical appliances

To size a grid-connected photovoltaic installation, we first need to calculate the electricity consumption in kWh.

The following elements need to be identified:

- the power (in watts) of the electrical appliance
- duration of operation per day
- the number of operating days per year.

To calculate the annual electricity consumption CAE (kWh), the number of unit hours of use per day H_{uu} is multiplied by the number of days of use J_u and the power of the appliance P_a in W. This product is divided by 1000, as shown in the following equation:

$$CAE\ (kWh) = \frac{((H_{uu}\ J_u\ P_a))}{1000} \quad (2.1)$$

With :

H_{uu}: designates the unit hours of use per day

J_u: designates the days of use per year

P_u: designates the unit power in watts (W)

P_a: designates the total power of the appliance in watts (W)

Nr: refers to the number of items of equipment.

Tables 2.3, 2.4, 2.5, 2.6 and 2.7 show all the electrical requirements for the mechanical department.

Table 2.3. Electrical requirements for the administration room

Equipment	Nr	H_{uu} (h)	J_u (days/year)	P_u(W)	P_a(W)	CAE (kWh)
Computers	2	8	217	250	500	868
Tubes	6	12	217	40	240	624.96
air conditioners	1	10	67	3500	3500	2345
Printer	2	1	217	650	1300	282.1
Refrigerator	1	5	67	500	500	167.5
Total	12	36	785	4940	6040	3213

Table 2.4. Electrical requirements of the laboratory

Equipment	Nr	H_{uu}(h)	J_u (days/year)	P_u (W)	P_a (W)	CAE (kWh)
Computers	4	8	267	250	1000	2136
Tubes	6	12	267	40	240	768.96
air conditioners	1	10	79	3500	3500	2765
Video projector	1	5	40	624	624	124.8
Printer	2	1	267	650	650	173.55
Total	13	36	920	5064	6014	5968.31

Table 2.5. Electrical requirements of a teaching room

Equipment	Nr	Huu (h)	Ju (days/year)	Pu(W)	Pa (W)	CAE (kWh)
Computers	1	8	267	250	250	534
Tubes	6	12	267	40	240	768.96
Air conditioners	1	10	67	3500	3500	2345
Video projectors	1	5	267	624	624	833.04
Total	9	35	868	4414	4614	2879

Table 2.5. Electrical requirements of a computer room

Equipment	Nr	Huu (h)	Ju (days/year)	Pu(W)	Pa(W)	CAE (kWh)
Computers	7	8	267	250	1750	3738
Tubes	6	12	267	40	240	768.96
air conditioners	1	10	67	3500	3500	2345
Video projectors	1	5	267	624	624	833.04è
Total	15	35	868	4414	6114	6083

Table 2.6. Electrical requirements of a teaching office

Equipment	Nr	Huu (h)	Ju (days/year)	Pu (W)	Pa (W)	CAE (kWh
Computers	1	8	267	250	250	534
Tubes	4	12	267	40	160	512.64
air conditioners	1	10	67	3500	3500	2345
Printer	1	1	267	650	650	173.55
Total	7	31	868	4440	4560	2879

Table 2.7. Electrical requirements of the sanitary block

Equipment	Nr	Huu (h)	Ju (days/year)	Pu (W)	Pa (W)	CAE (kWh)
Lamp	4	10	267	40	160	427.2

The electrical requirements of the Mechanical Engineering department can therefore be summarised in table (2.8).

The total annual electricity consumption CAET for the Mechanical Engineering department is calculated using the following formula:

$$C_{AET} = N_r C_{AEU} \quad (2.2)$$

With :

Nr: refers to the number of rooms.

CAEU: refers to the annual unit electricity consumption calculated in the previous tables.

Table 2.8. Electrical requirements of the mechanical engineering department

Rooms	Nr	Pu (W)	Pa (W)	CAEU (W)	CAET (kWh)
Administration room	1	6040	6040	5213	5213

Laboratories	4	6014	24056	5968.31	23873.24
Computer rooms	6	4614	27684	6083	36498
Teaching rooms	9	6114	55026	2879	25911
Teaching offices	6	4560	27360	2879	17274
Toilets	10	427.2	4272	160	1600
Total	36	27342	144438	17214	86496

According to the calculation of the electrical requirements of the mechanical engineering department summarised in table 2.8, the electricity consumption in kWh is equal to 86496 kWh.

3.2 Choice of solutions

Grid-connected rooftop photovoltaic systems generally face south. However, vertical systems installed on facades also use east-west radiation. This shifts production to the morning or afternoon hours and dampens the mid-day peaks.

In our work, we propose two solutions for integrating photovoltaic energy into the Mechanical Engineering department. The first solution consists of solar panels on the roof or integrated into the building. The second solution consists of installing solar panels on the facades, if the available surface area is not exceeded.

4. Calculating and sizing a photovoltaic installation on the roof

The second step in sizing a medium-voltage photovoltaic system is to calculate the system's peak power, which is used to select all the components needed to install the photovoltaic system.

4.1 The peak power of the photovoltaic system

The rated power of a PV module is called its peak power. It is expressed in Wp (Watt-peak) or kWp (kilowatt-peak). In fact, the energy consumed per year is equal to kWh/year and STEG tolerates a specific production of around 1kW$_c$for each 1800 kWh/year of the adapted site. The peak power of the installation Pc in kWp is calculated by the following equation:

$$P_c = \frac{P_{com}}{P_{sp}} \qquad (2.3)$$

With :

P_{com}: Annual consumption power equal to 86496 kWh/year

P_{sp}: Specific annual production for 1 kWp in kWh/year equal to 1800 kWh/year

According to the calculations, the peak power of the installation in question is equal to Pc = 48.053 kWp.

In this application, to avoid the risks of overvoltage and circuit protection, we will increase the value of the peak power PC to 50 kWp.

4.2 Components of the photovoltaic system

4.2.1 Photovoltaic panel

The solar module used in our study is from Canadian Solar Inc. and has a peak power of P_c= 400 W_c. It is characterised by an open-circuit voltage V_{oc} = of 53.4 V and a short-circuit current equal to I_{sc}= 9.60 A. The ABB 50 kW delivers a
voltage of V_{mpp}= 44.1 V while delivering a current of I_{mpp}= 9.08 A at the point of maximum power. This module complies with the NF EN 61730 series of standards.

Figure 2.2. Canadian Solar Inc. photovoltaic panel.

Table 2.9 describes the necessary characteristics of the Canadian Solar Inc. mono-crystalline silicon solar panel that we have chosen for our installation. Other parameters are also shown in the data sheet attached in Appendix 1.

Table 2.9. Technical characteristics of the Canadian Solar Inc.

Module reference	CS1U - 400MS	
Maximum power (P_{max})	P_{mpp}	400 WP
Voltage at P_{max}	Vmp	44.1 V
Open circuit voltage	Voc	53.4 V
Current at P_{max}	Imp	9.08 A
Short circuit current	Isc	9.60 A
Panel length	L	2.078 m
Panel width	L	0.992

4.2.2 Photovoltaic inverter

The function of the inverter is to convert the electricity produced as efficiently and safely as possible to the electricity grid. It takes the form of a metal box fitted with a radiator or fan. It is placed on a vertical support or in an electrical cabinet. The inverter converts the direct current from the photovoltaic modules into alternating current identical to that from the grid. In its conversion, the inverter seeks the maximum point of function (MPP) at each moment, according to the I/U characteristics of the photovoltaic field.

Our choice of inverter is the ABB TRIO-TM-50_0-400 model, an inverter with 3 independent MPPTs and a power rating of up to 50 kW and a three-phase connection for decentralised photovoltaic installations. Modular and economical, this inverter is designed for rooftop installations.

Figure 2.3. ABB 50 kW inverter with 3 MPPT

Table 2.10 describes the overall characteristics of the inverters used, as well as the electrical parameters of the inverter. In appendix 2, we can refer to the rest of its characteristics presented in its technical data sheet.

Table 2.10. Technical characteristics of the ABB type inverter

Electrical parameters of the inverter	
Inverter reference	TRIO-TM-50_0-400
DC	
Max DC power	17500 W
Max DC voltage	1000 V
DC start input voltage Vstart	420...700 V (Default 420 V)
MPP voltage range	0.7xV start ...950 V (min 300 V)
Max input current per MPPT	36 A
Number of MPPTs	3
AC	
Output power	50000 W
Max output power	50000 W
Max output current	77 A
Voltage range	320...480 V
Operating frequency	50Hz
Power factor	> 99.5 %
Mains connection	Three-phase
System	
Euro efficiency	98.0 %
Degree of protection	IP65

The DC-AC converter can be sized according to the following criteria:

- Input voltage
- Output voltage
- Maximum DC power
- Maximum AC power

4.2.3 Installation configuration

To determine the number of Ppv panels required for this installation, the ratio of the total peak power to the maximum power of the panel must be calculated according to the following equation:

$$N_{pv} = \frac{P_c}{P_{pv}} \qquad (2.4)$$

With :

P_{pv}: Maximum power of a panel

P_c: Peak power

According to our calculation, Npv = 125 panels. By increasing this value, we choose 126 panels.

In our installation, we chose to use the ABB inverter and 126 Canadian Solar Inc. model CS1U - 400MS photovoltaic panels.

The minimum $^{(,,)}$ $N_{(min)\ (>s)}$" and maximum $^{(,,)}$ $N_{(max)\ (>s)}$" number of photovoltaic modules in series compatible with the MPPT voltage range of the inverter are expressed by the following equations:

$$N_{max,s} = PE_I\left(\frac{U_{mppt,max}}{U_{oc(-10°c)}}\right) \quad (2.5)$$

$$N_{min,s} = PE_S\left(\frac{U_{mppt,min}}{U_{mpp(80°c)}}\right) \quad (2.6)$$

With :

PEI: Lower integer part.

PES: Upper integer part.

U_{mp}pt,min: Minimum value of the tracker voltage (V).

U_{mp}pt,max: Maximum value of the tracker voltage (V).

The value of the open-circuit voltage of the photovoltaic modules $U_{(OC)\ (\_(10)\ "C)}$ (V) at a minimum temperature of Tmin = -10°C is calculated by the equation :

$$U_{oc(-10°c)} = K_u\ U_{oc(STC)} \quad (2.7)$$

Where Ku is the correction factor taking into account the increase in the open circuit voltage of the modules.

Considering the minimum ambient temperature Tmin = -10°C of the PV installation site and the temperature coefficient of the open-circuit voltage equal to a_{Uoc}= -0.29 %, the correlation factor Ku is calculated by the following formula:

$$Ku = 1 + \left(\frac{\alpha_{Uoc}}{100}\right)(T_{min} - 25) \quad (2.8)$$

In our application, we find a correlation factor Ku = 1.1015.

We can therefore deduce that the value of the open-circuit voltage of the photovoltaic modules is equal to $U_{(oc()\ \_(10)\ oc)}$= 59.0V.

Similarly, we can calculate the value of the MPP point voltage of the photovoltaic modules $U_{(mp)\ p}(80^{(o)\ c})$ *(V)* at a maximum temperature equal to T_{max}= 80°C, calculated by the following equation:

$$U_{mpp(80°C)} = Ku\ U_{mp} \quad (2.9)$$

Under these conditions, the correction factor Ku takes into account the decrease in the closed-circuit voltage of the modules. Considering the maximum ambient temperature $T_{max}$of the photovoltaic installation site and the maximum power temperature coefficient equal to $\alpha_{Ump} = -0.37\,\%$, , the correlation factor Ku is calculated using the following formula:

$$Ku = 1 + \left(\frac{\alpha_{Ump}}{100}\right)(T_{max} - 25) \quad (2.10)$$

Under these calculation conditions, the correlation factor is Ku = 0.79.

Furthermore, the maximum power voltage of the photovoltaic modules at a temperature equal to T_{max}= 80°C is calculated using the following formula:

$$U_{mpp(80°C)} = Ku\ U_{mp} \quad (2.11)$$

Under these conditions, the maximum power voltage of the photovoltaic modules is equal to $U_{(mpp(80)\ oC)}$= 34.8V.

According to equations 2.4 and 2.5, in our case this is :

$N_{maX;S}$= 16 panels

$N_{min>s}$= 8 panels

The number of strings of photovoltaic modules in parallel compatible with the DC input current is expressed by the equation we find :

$$N_{max,p} = PE_I\left(\frac{I_{max}}{I_{mpp}}\right) \quad (2.12)$$

So $N_{max,p} = \frac{36}{9.08} = 3$ strings

With :

PEI: Lower integer part,

$N_{max;P}$: Maximum number of strings in parallel,

I_{max}: Maximum current allowed by MPPT of the inverter (A),

I_{mpp}: Maximum module power current (A).

In this installation, we have chosen an ABB trio inverter with 3 MPPTs. Each MPPT can accept 3 strings.

The number of panels per chain $Nb_{(p)/c}h$ is calculated using the following formula:

$$N_{bp/ch} = N_{bp}\ N_{bch} \quad (2.13)$$

With :

Nbp: Total number of panels equal to 126 panels

Nbch: Number of strings equal to 9 strings

We have therefore chosen an installation of 9 strings in parallel. Each string contains Nbp/ch equal to 14 panels in series.

4.2.4 Inverter and PV array compatibility

4.2.4.1 Voltage compatibility

The voltage level across the string inverter results from the sum of the voltages of the modules connected in series in the string. The voltage across the microinverter corresponds to the voltage of the module to which it is connected. The voltage of the photovoltaic module and the voltage of the photovoltaic generator are :

- Greatly affected by temperature, as the voltage decreases considerably as the temperature rises.
- Only slightly affected by solar radiation, as the voltage rises very quickly and then remains relatively constant in relation to radiation.

The maximum voltage of the modules in a photovoltaic string is calculated in winter conditions at a temperature of -10° C. It must be less than the maximum input voltage of the inverter.

The minimum module voltage in a photovoltaic string is calculated for a module temperature of up to 80 °C. It must be higher than the minimum value of the MPPT voltage range of the inverter.

Voltage compatibility for lows at -10°C verifies the following inequality condition:

$$14\ U_{oc(-10°C)} < U_{max\ de\ l'onduleur} \quad (2.14)$$

This condition is verified knowing that: 14x 59.0 = 826V < 1000V

Voltage compatibility for high temperatures at 80°C enables the following inequality condition to be verified.

$$14\ U_{mpp(80°C)} > U_{mpp,min} \quad (2.15)$$

This condition is verified knowing that: 14 x 34.8 = 487.2 *V* > 300 *V*.

4.2.4.2 Current compatibility

The value of the DC current at the inverter input is :

- Strongly affected by solar radiation (direct linear relationship) ;

- Slightly affected by temperature (the current increases slightly as the temperature rises).

The maximum current $I_{pv\ max}$delivered by the photovoltaic generator must not exceed the maximum inverter input current $I_{eo\ max}$; i.e. :

$$I_{pv\ max} \leq I_{eo\ max} \quad (2.16)$$

The maximum current of the photovoltaic generator is determined by the number of strings strings connected in parallel according to this equation :

$$I_{sc\ (80°C)} = \left((T_{max} - 25) \times \frac{\alpha}{100} \times N_{max,p}\right) + I_{sc} \quad (2.17)$$

$I_{(sc)}\ (80^{(o)\ c}) = 9.85$ A.

With :

T_{max}: Maximum module temperature taken as 80°C

a: current/temperature coefficient of the photovoltaic module given by the module manufacturer in mA/°C is equal to 0.05 % / °C.

$N_{(max),p}$: Number of strings of photovoltaic modules in parallel equal to 9 strings.

I_{sc}: Open circuit current of the photovoltaic modules is equal to I_{sc}= 9.60 A.

maximum current= number of strings in parallel x I_{mpp}x 1.25

maximum current = 9 x 9.08 x 1.25 = 102.15 A < 108 A.

4.2.4.3 Power compatibility

The powers of the photovoltaic generator and the inverter must be mutually matched. The nominal AC power of the inverters can be equal to the power of the photovoltaic generator (in STC conditions) with a tolerance of ±10%. In these conditions, we write

$$0.8 \times P_n < P_{AC\ no} < 1.2 \times P_n \quad (2.18)$$

where

P_n: Nominal power of the photovoltaic generator

PAC no: Nominal AC power of the inverter.

This inequation is verified by knowing that :

0.8x50000 = 40000 W < 50000 W < 1.2x50000 =60000 W

The photovoltaic panels in the installation are therefore compatible with the chosen inverter.

4.3 Setting up electrical safety tools for the installation

Figure (2.4) shows the electrical safety equipment recommended for a photovoltaic installation.

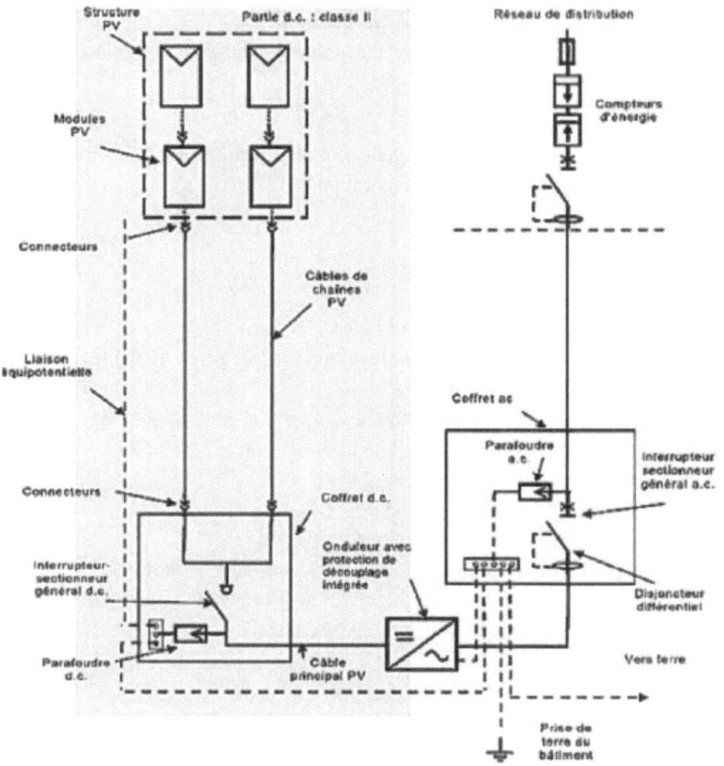

Figure 2.4. Recommended electrical safety equipment for a photovoltaic installation.

4.3.1 Connectors

All the connection points required to make a PV string with a *maximum* voltage $U_{(oc)}$) greater than 60 V (two or more modules) must be provided with connectors, including at the ends. To guarantee the quality of the connection and limit the risk of arcing, each pair of male-female connectors to be assembled must be of the same type and brand.

In our installation, we used DC MC4 6 mm^2type connectors with a rated current of 30 A, a rated voltage of 1000 V, protection class II and an IP65 environmental protection rating (see Appendix 3).

Figure 2.5. Mc4 connector

4.3.2 DC cabinet

All the DC protection devices (isolators, surge arresters) are often grouped together in a single cabinet located at the inverter input (DC cabinet). In this installation, we will be using three watertight 3-string / 1 MPPT DC cabinets for the 126 modules, with IP 65 protection from Eutélec.

4.3.2.1 Choice of fuses

In an installation with several PV module strings in parallel, the modules must be protected against the effect of reverse currents that may originate in the faulted strings.

- If the PV generator consists of a single string, there is no reverse fault current and no overcurrent protection is required.
- If the PV generator consists of two strings in parallel, the maximum reverse current that can flow in the faulted string can be up to $I_{SC}max$.

The modules in the faulted string are still able to withstand this reverse current and no string overcurrent protection is required.

- If the PV generator is made up of $N_c$strings in parallel with each MPPT $(N_c > 2)$, then the maximum reverse current that can flow in the faulted string can be up to $(N_c - 1)$ Iscmax.

A string protection device against overcurrents is only required if the number of strings in the generator $N_c$is greater than $N_{(cmax)}$ (the maximum number of unprotected parallel strings for a PV generator) and is expressed by the following formula:

$$\mathbf{N_{cmax} \leq \left(1 + \frac{I_{rm}}{I_{sc(STC)}}\right)} \qquad (2.19)$$

With :

$I_{(SC)}(STC)$: PV module short-circuit current at STC conditions.

I_{rm}: Maximum reverse current of the solar module.

In our study we have I_{rm}= 15 A, $I_{(SC)}(STC)$ = 9.60 A, N_c= 9. Thus we find $N_{cmax} \cup 2$.

The maximum number of parallel strings (N_{pmax}) per protection device is expressed by the following formula:

$$\mathbf{N_{pmax} \leq 0.5\left(1 + \frac{I_{rm}}{I_{sc(STC)}}\right)} \qquad (2.20)$$

With :

I_{rm}= 15 A, $I_{(sc)\,(()\,(S)}TC)$ = 9.60 A, N_c= 9.

According to these calculations, we find N_{pmax}= 1, so we protect each chain with a single protective device. Under these conditions, the following inequality condition must be met:

$$\mathbf{1.1\, I_{smax} \leq I_n \leq I_{rm}} \qquad (2.21)$$

This inequality is verified knowing that 1.1 I_{smax}=13.2 A and $I_{(rm)}$) = 15 A.

Therefore: 13.2A < I_n< 15 A

We therefore choose the rated current of the protection devices I_n=14 A as the chain fuse rating.

The overcurrent protection devices must be either fuses conforming to standard NF EN 60269-6, or circuit breakers conforming to the same standard.

In our installation, the number of generator strings $N_c$is less than 20.

Therefore, the conventional operating current I_z for the circuit-breakers is taken to be equal to 1.3 I_n, whereas for the fuses, the current I_z is taken to be equal to 1.45 I_n.
In these conditions, we have chosen fuse protection because it is the device most commonly used for photovoltaic installations.

$$I_z \geq 1.45\, In \qquad (2.22)$$

Given that I_n= 15 A, we need to check that :
I_z> 21.75 A
To select the DC side fuse criteria, check that :

- The rated operating voltage U_e must be greater than or equal to the PV field voltage U_{ocmax} calculated by the following formula:

$$U_{oc\,max} = k\, U_{oc\,pv}\, x\, N_p \qquad (2.23)$$

With : κ: Correction factor equal to 1.2.
Uoc_{PV}: PV field voltage equal to 53.4 V.
Np: Number of panels in series equal to 14.
Thus, we find that Uoc_{max} is equal to 897.112 V.
Referring to the rated current I_n, the rated breaking capacity P_c must be at least equal to Iscmax of the PV field and the marking G_{pv} in accordance with NF en 60269-6.
In our case, and according to the characteristics of the solar panels, the maximum reverse current of the panel used is equal to 15 A. The protection device is therefore necessary. In these conditions, we propose using 6 CH10x38 gPV 14A/1000V DC fuses to protect the photovoltaic systems against overloads and short-circuits. This choice complies with the standards and directives of IEC 60269-6 /UL standard ML E358455 (see Appendix 4).

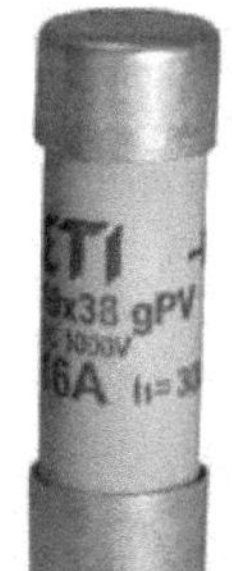

Figure 2.6. DC fuse.
The characteristics of the fuse are summarised in table 2.11.
Table 2.11. Fuse characteristics

Model	CH10x38 gPV
Symbol	002625079
Size	10x38 mm
Rated current	30 kA
Voltage rating	1000CL/R=2ms

4.3.2.2 Selecting the isolating switch
To enable maintenance of PV inverters and PV modules, disconnecting means must be

provided close to the DC side of the inverter. The following criteria should be taken into account when selecting the isolating switch:

- Rated operating voltage DC (U_e) greater than the maximum short-circuit voltage of the panels per string ($U_{comax/ch}$), i.e. :

$U_e > U_{comax/ch}$ (2.24)

- Rated operating current $I_e$greater than the maximum service current I_{ms}i.e. :

$I_e > I_{ms}$ (2.25)

Chain 1 disconnector

In the chain 1 disconnector, we have :

$U_{comax/ch} = 826$ V

$I_{ms} = 1.25$ Icc = 1.25^9.60 =12 A

In this installation, all 9 chains are identical. We therefore choose the disconnector with the same maximum short-circuit voltage and the same service current.

The right choice for this installation is three identical ZJ Benny 4P 1200V 63A DC disconnectors, type BB1-63, with the following characteristics (see Appendix 5):

- Rated operating voltage U_e= 1200 VDC
- Rated operating current I_e= 32 A

Figure 2.7. DC switch-disconnector

4.3.2.3 Choice of DC surge arrester

DC surge arrester chain 1

The choice of DC arrester is based on the following criteria:

- $U_{cpv} > U_{oc\,max}$, with $U_{ocMAX} = 826$ V
- $U_{cpv} > U_{ocSTCmax}$
- Up: Lightning arrester protection level, this value must be less than or equal to 80% of the surge withstand voltage U_{imp} of the electrical equipment to be protected (Up<0.8 U_{imp})
- $I_{max} = 40$ kA
- $I_n > 5$ kA
- $Icc_{pv} > 1.25 \times I_{scSTC}$ of the PV generator.

With :

U_{cpv}: maximum steady state voltage

U_{ocmax}: maximum photovoltaic generator voltage

Uoc STC = Np X V_{oc}

I_{max}: Maximum discharge current

I_n : Rated discharge current
U_p: Level of protection
$I_{cc\,pv}$: Short-circuit current withstand of an arrester
U_{imp}: Rated impulse withstand voltage = 6.480 kV (according to guide UTE C 15-712-1).

Verification :

- $U_{cpv} > U_{ocMAx}$ with UOCMAX at -10°C/sheath = 826 V
- Ucpv(v) > $U_{oc\ STC\ max}$

with :
Uoc STC max = 1.2 x Uoc STC= 897.12 V
(2.26)
$U_{oc\ STC}$= 747.6 V

- Up < 0.8 X Uimp

Up < 0.8 X 6.480 kV = 5.184 kV

- Imax = 40 kA
- In > 5kA.

The choice of DC arrester for all the remaining chains is based on the same characteristics as the first chain.

On the basis of the above criteria, 2 type 2 SPD-DC MZ- TYPE 3 DC surge protectors were chosen for each enclosure, made by DIGITAL ELECTRIC (see appendix 6).each surge protector has the following characteristics:

- U_{cpv}= 1000 V: maximum steady state voltage
- I_n = 20 kA: nominal discharge current
- $U_{(p)}$ <3.2 kV: protection level
- I_{max}= 40 kA: maximum discharge current.

Figure 2.8. DC surge arrester

4.3.3 Choice of DC cables

Cable cross-sections must be carefully calculated to limit line losses (maximum 3% for all cables).

It is recommended that PV installations use cables with a cross-section of 6 mm^2. We will now check that this cross-section is sufficient for our installation, based on two conditions:

- The admissible current.
- The maximum voltage drop

4.3.3.1 Voltage drop

For string 1 :

The voltage drop across the DC cable is as follows:

$\Delta U(V) = b\ R\ I$

(2.27)

$$\Delta U(\%) = \frac{\Delta U \times 100}{U} \qquad (2\text{-}28)$$

With: b = 2 (in continuous operation a double loss phenomenon).
I: rated module current equal to = 9.08 A.
The resistance R is calculated from this formula:

$$R = \frac{\rho \times L}{S} = 0.15\ \Omega$$

With :
L: length between photovoltaic field and inverter [m] equal to 40 m p: conductivity of copper at a temperature of 90°C equal to $0.0225\ \Omega mm^2/m$
S: Conductor cross-section = 6 mm^2.
U : Nominal voltage of the module x number of modules = 44.1 x14 = 617.4 V

$$\Delta U(V) = 2.724\ V < 6.9\ V$$

$$\Delta U(\%) = 0.44\ \% < 3\%$$

The same calculation of the voltage drop across the DC cable is repeated for the 8 cams, since the 9 strings are identical.
The 6 mm^2cross-section of the DC cable is therefore sufficient for our installation, depending on the voltage drop. (See Appendix 7)

4.3.3.2 Current carrying capacity

The ampacity of a cable is the maximum value of the current that can permanently flow through the conductor without its temperature exceeding its specified temperature.

Table 2.12. Current-carrying capacity IZ of photovoltaic cables for a maximum core temperature of
of 90°C (extract from UTE guide 32-502).

Cross-section (mm^2)	Two adjacent cables on wall
2.5	19 A
4	26 A
6	33 A
10	46 A
16	62 A
25	82 A
35	102 A

We have IZ = 1.25 x ICC = 1.25 x 9.60 = 12 A
So, according to table 2.12, the 6 mm^2cross-section of the DC cable is sufficient for our installation.

4.3.4 AC cabinet

All the AC protection equipment (circuit-breaker, lightning arrester) is generally housed in a single enclosure located at the inverter output (AC enclosure). In this installation, we will be using a single enclosure with 126 watertight modules with IP 65 protection and ZJ Benny brand class.

4.3.4.1 Choosing the circuit-breaker

The circuit-breaker is an electromechanical device used to protect the electrical installation against overloads and short-circuits. Thanks to their differential section, which detects a difference in current intensity between phase and neutral, they cut off the current to protect people.

Normally high currents such as short-circuit, overload and indirect contact can occur in electrical installations. To protect against these, it is preferable to install an earth leakage circuit breaker (ELCB) to cut off the installation if necessary. On the AC side, an earth leakage circuit breaker should be installed upstream of each inverter. The choice of circuit-breaker in this installation is based on the following criteria:

Rated voltage Uc = 230/400 V

In > Is.

Where :

In: Circuit-breaker rated current.

Is: maximum inverter output current = 77 A.

We therefore use a single 4-pole circuit breaker, type C120N AC, Schneider, with a rating of 100 A, mains voltage U = 380 V, rated residual current I_n= 10 kA and sensitivity 300 mA. (See Appendix 8).

Figure 2.13. 4P earth leakage circuit breaker.

4.3.4.2 Choosing the AC surge arrester

To protect inverters against possible overvoltages from the distribution network, it is often necessary to provide a surge arrester with its own specific protection at generator voltage level.

The choice of AC surge protector is based on the following criteria:

In >> 5 kA

Up << U_{surge} = 4000 V (according to guide UTE C 15-712-1)

Imax >> 18 kA.

Therefore, 4 Digital Electric type JVS 16-C and type 2 three-phase AC surge protectors (see appendix 9) with the following characteristics must be used for the single enclosure:

- Us : Operating voltage Us = 320 V
- In = 20kA : Rated discharge current
- Up < 1.5 kV Protection level
- Imax = 40 kA Maximum discharge current.

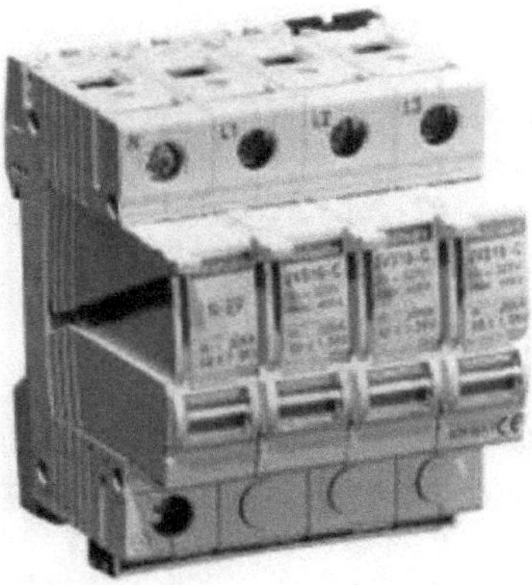

Figure 2.14. Three-phase AC arrester.

4.3.5 Choice of AC cables

For photovoltaic installations directly connected to the public distribution network, the maximum voltage drop permitted between the AC terminals of the inverter and the connection circuit breaker is 3% at the rated power of the inverter(s). It is recommended to use cables with a cross-section of 16 mm²in PV installations. We should therefore check that this cross-section is sufficient for our installation.

The voltage drop across the AC cable at 90°C is calculated using the following formula:

$$\Delta U = b\,(\frac{\rho \times L}{s} \times COS\O + \lambda \times L \times SIN\O) \times I \qquad (2.28)$$

With :

L: single conductor length [m] to STEG meter =50 m

I: current per phase equal to 77A

P: copper conductivity at 90°C equal to 0.0225 mm^2/m

S : conductor cross-section in [mm2] equal to 6 mm2

U : three-phase voltage equal to 400 V b = 1 (three-phase operation)

$SIN\O = 0$

$COS\O = 1$

Cable reactance λ equal to 1

Thus, we find $\Delta U = 5.41 V < 6.9\ V$

$\Delta U(\%) = \frac{\Delta U \times 100}{U} (\%) = 1.35\% < 3\ \%$

The 16 mm2 cross-section of the AC cable is therefore sufficient for our installation. (See Appendix 10)

4.4 Overvoltage protection description

To minimise the effects of induced overvoltages, the metal structures of the modules and the metal support structures must be connected to a protective equipotential bonding system which is itself connected to earth.

In our installation, the cross-section of the earthing cable from the equipotential bonding strip to the earth connection is 25 mm2.

4.4.1 Earthing the grounding and conducting elements of the DC section

Each structure is earthed via an earth cable with a cross-section of 16 mm2 using earth splitters. Earthing between the panels is also carried out using earth cables with a cross-section of 6 mm2.

4.4.2 Earthing the AC part of the installation

All the grounds on the AC side must be earthed by a protective conductor.

The inverter is earthed via a 16 mm2 earth cable.

The structure of the photovoltaic system against the wind for 126 panels must be validated by the control offices.

5. Integration of photovoltaic systems on facades

In this work, the photovoltaic system is connected to the grid and integrated on the facades of vertical buildings. For decades, architects have been asking for solutions to customise the colour of photovoltaic elements, making them easier to integrate into buildings. In this area of BIPV, additional standards and current building codes have to be complied with.

For our purposes, the power output of a grid-connected rooftop photovoltaic system is 50.0 kWp. This rises to 75 kWp on the facades of the mechanical engineering department. The PVSYST professional photovoltaic production simulation software is used to calculate the dimensions.

Given that the surface area on the facades of the department is insufficient to integrate photovoltaic systems, this solution is not feasible.

Thus, the integration of photovoltaic systems for vertical buildings on facades is a rejected solution and the only solution is to integrate the photovoltaic modules on the roof.

6. Signage and labelling on photovoltaic system components

For safety reasons, it is imperative that all cables are labelled and the presence of a photovoltaic system on a building is indicated to the various parties involved (maintenance personnel, inspectors, public distribution network operator, emergency services).

6.1 Labelling the DC part

All junction boxes (PV generator - PV group) and DC lines must bear a visible and indelible marking indicating that active parts inside these boxes may remain live even after the DC side of the inverter has been disconnected.

The label bearing the words "Caution live DC cable" must be affixed :

- On the front of the junction box
- On the front of the DC cabinets
- On the ends of the DC cables

Figure 2.15. "Caution live DC cable" label

The "Do not operate on load" label must be fitted :

- Inside DC junction boxes and enclosures
- Near fuse-disconnectors and surge arresters

Figure 2.16. Do not operate on load" label

6.2 Labelling on the inverter

All inverters must bear a marking indicating that before any work is carried out, the 2 voltage sources must be isolated.

Figure 2.17. Labelling on the inverter

6.3 Labelling of the AC part

A label located near the device ensuring the concession limit: Connection circuit breaker (if power limited) or low-voltage switchboard (if power monitored).

Figure 2.18. Distribution network outage

In conclusion, our roof-integrated PV installation requires a photovoltaic array consisting of a 50.0 kWp system. This system consists of 126 monocrystalline modules of 400 $W_c$for a total of 50,000 W_c. All the modules will be connected to an ABB 50 kWp inverter to supply AC voltage. The system also includes protection boxes (three DC protection boxes and an AC protection box), to guarantee electrical protection on the photovoltaic generator side on the one hand and on the grid side on the other, in accordance with the standards for photovoltaic installations and the specifications of STEG and ANME for grid connections. This installation is made up of the elements listed in table 2.13.

Table 2.13. Main components of a 50.0 kWp installation

Description	Number	Brand	Reference	Appendix
Modules	126	Canadian Solar Inc.	HiDM CS1U-400MS	Appendix 1
Inverter	1	ABB	TRIO-TM-50.0-400	Appendix 2
Isolating switch DC	3	ZJ Benny	BB1-63	Annex 5
Surge arrester DC	6	Digital Electric	08404 Surge Protector DC 1000V	Annex 6
Fuse	6	Eti group	002625079 CH10x38 gPV 14A/1000V Dc	Annex 4
Surge arrester AC	4	Digital Electric	Tetra 08204 JVS 16-C	Appendix 9
Earth leakage circuit breaker	1	Schneider	A9N18374 C120N circuit breaker 4P 100A	Appendix 8
DC cable	LOT	KBE Solar	H1Z2Z2-K Section 6 mm^2	Appendix 7
AC cable	LOT	DestocKable	U1000R2V	Appendix 10
Connector	LOT	Digital Electric	MC4 41312-41314	Appendix 3

7. Technical and economic study of a photovoltaic installation

The cost of a photovoltaic installation comes mainly from the purchase price of the photovoltaic modules. However, prices can vary greatly depending on the type of architectural integration chosen, the choice of structure (fixed or mobile), the connection to the grid and the size of the installation.

Table 2.14 illustrates the prices of the various components of a grid-connected rooftop photovoltaic system.

Table 2.14. Cost estimate for a 50.0 kWp system

Number	Description	Qty	VAT	PU	PTHT	INCL. VAT (DT)
1	Photovoltaic module 400 Wp	126	7%	700 000	88 200 000	94 374 000
2	ABB inverter 50.0	1	7%	25 000 000	25 000 000	26 750 000
3	DC protection box	3	19%	650 000	1 950 000	2 320 500
4	AC protection box	1	19%	550 000	550 000	654 500
5	Galvanised metal structure	1	19%	13 000 000	13 000 000	15 470 000

6	Wiring and earthing	1	19%	4 000 000	4 000 000	4 760 000
7	Installation costs	1	19%	5 000 000	5 000 000	5 950 000
8	Stamp					0.600
Total				48 900 000	137 700 000	150 279 600

Including :

Qty	Quantity of photovoltaic components
VAT	Value added tax
PU	Unit price of components
PTHT	Total price before tax
INCLUDING VAT	Total including tax

8. Simulation and design of our installation using PVsyst software

In our study and design of this PV installation, we will use the PVsyst 6.8.1 software:

8.1 Description of PVsyst

PVsyst is designed to be used by architects, engineers and researchers, but it is also a very useful educational tool. It includes extensive context-sensitive help, which explains in detail the procedure and models used, and offers an economic approach to guide project development.

PVSYST is a solar panel sizing software package that provides a range of information, such as energy production, irradiation and installation costs, surface area required and annual energy production. An advanced mode allows you to obtain much more information for a very complete study.

8.2 Simulation

We simulated our photovoltaic installation on the roof of the mechanical engineering department with a power of 50 kWp and on the facades with a power of 75 kWp using PVsyst software version 6.8.1.

8.2.1 Grid-connected photovoltaic system 50 kWp

The first step in sizing the PVsyst photovoltaic system is to manage the weather databases and the system components.

For our large-scale grid-connected photovoltaic installation, the geographical site is located in the southern region, government of Sfax at the Ecole Nationale d'Ingénieurs, as shown in Figure 2.19 (see Annex 11).

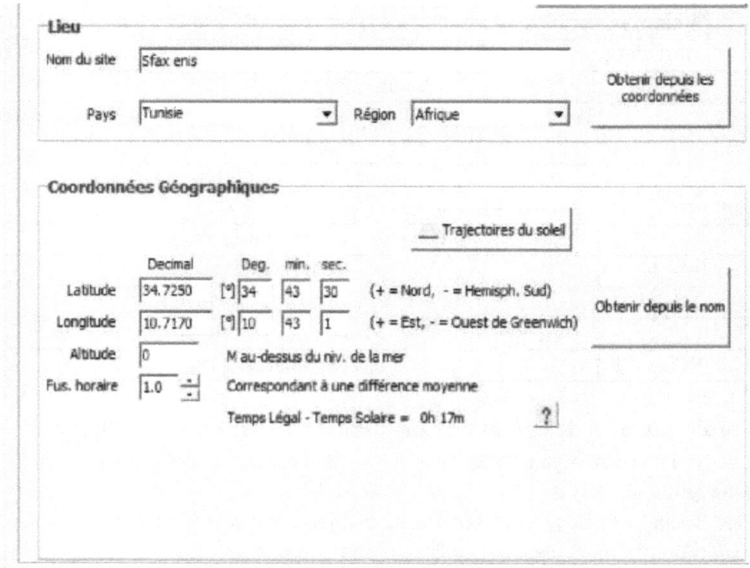

Figure 2.19. Geographical coordinates of the ENIS Sfax site.

In the second stage, we studied and analysed the project in detail for a grid-connected system, defining the main input parameters such as orientation. For the 50 kWp photovoltaic system, the type of PV field is a fixed slope with a 30° tilt and southern orientation.

Our grid-connected rooftop photovoltaic installation requires configuring the choice of panel and inverter compatible with the 50 kWp peak power as shown in the following figure 2.20.

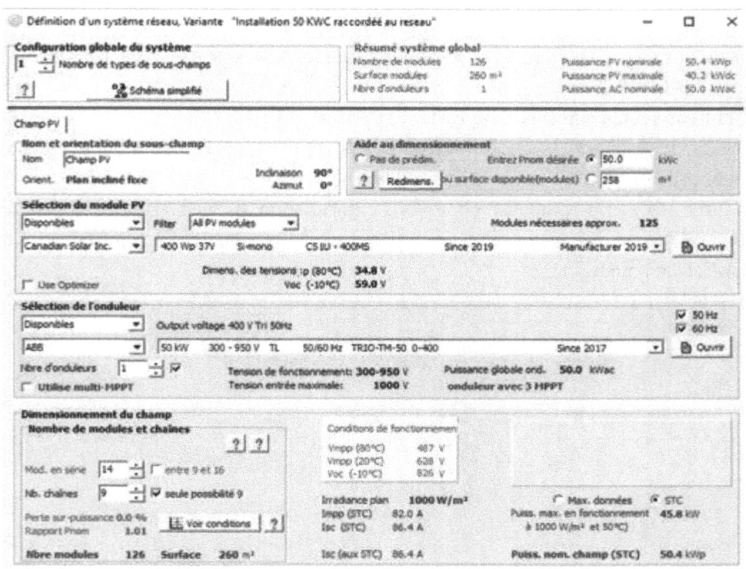

Figure 2.20. Grid system.

8.3 Grid-connected photovoltaic system 75 kWp

The difference between the grid-connected photovoltaic installation on the roof 50 kWp and the photovoltaic installation on the facades 75 kWp is the change of orientation of the photovoltaic field which is presented in figure 2.21. (See Annex 12)

Figure 2.20. Panel orientation.

The dimensioning of this photovoltaic installation is based on the choice of a panel and an inverter that is compatible with the peak power of 75 kWp.

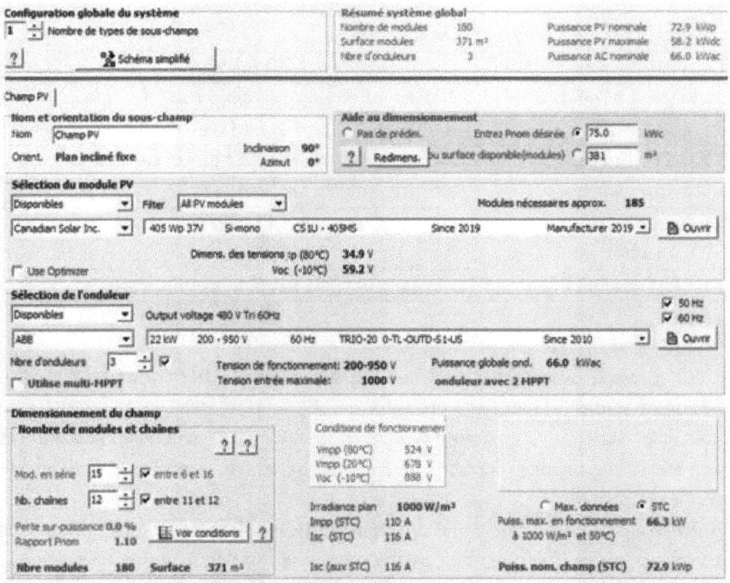

Figure 2.21. Sizing of the PV array.

9. Conclusion

In this chapter, we have carried out a theoretical study of our large-scale grid-connected photovoltaic installation. Then we presented the various elements making up the photovoltaic installation system, namely the PV panel, the inverter, the DC and AC cabinets and the electrical cables. Thus, the dimensioning of each element of the photovoltaic chain of a photovoltaic installation integrated on the roof of the mechanical engineering department at ENIS Sfax and we validated our dimensioning by the PVsyst software. In addition, a technico-economic study has been carried out which allows us to estimate the cost of the photovoltaic system, which is equal to 150,279,600 DT. In the following chapter, we will study and size a small-scale grid-connected photovoltaic installation.

Chapter 3

Study and analytical sizing of a small-scale small-scale photovoltaic installation

1. Introduction

In this chapter, we develop a study for the dimensioning of a photovoltaic installation connected to the LV grid on a small scale. We choose the LASEM laboratory as a model. First, we will calculate the electrical requirements. Then we calculate the electricity consumption. Next, we carry out the sizing for a grid-connected PV installation integrated on the roof and facades analytically, and validate our sizing with software for the photovoltaic system. We conclude this chapter with an economic study.

2. Identification of the LASEM laboratory

The details of the Electromechanical Systems Laboratory (LASEM) are summarised in Table 3.1.

Table 3.1: Identification of the LASEM laboratory

Laboratory name	LASEM
Supervising Ministry	Ministry of Higher Education and Scientific Research
University	Sfax
Establishment	Sfax National Engineering School
Laboratory address	Route de Soukra km 3, BP, 1173, 3038, Sfax, Tunisia
Laboratory identification number	LR 99ES36
Field of research	Mechanical Engineering

2.1 Presentation of the laboratory

LASEM is a research laboratory listed among the national laboratories of the Ministry of Higher Education and the State Secretariat for Scientific Research. Although the LASEM is based at the Sfax National Engineering School, it is made up of teacher-researchers from the ENIS mechanical engineering department and researchers in mechanics and physics from other universities.

LASEM's members are specialists in different and complementary disciplines (materials, energy, dynamics), and its research involves national projects within the framework of federative research projects and international cooperation activities with laboratories of common interest in various countries such as the United States, Canada, France and Germany. This policy fosters the diversity of themes and the influence of both ENIS and LASEM on a national scale towards the local and international industrial environment.

2.2 History of the laboratory

The laboratory was created in 1999 by grouping together 3 research units specialising in complementary areas of mechanical engineering. The units were brought together around an initial project involving the contribution of the three teams to the design, modelling and optimisation of electromechanical systems in general and transport systems in particular. From the outset, the laboratory was able to integrate most of the teacher-researchers from the mechanical engineering department and other flats at the University of Sfax, the University of the Centre and the University of Tunis.

Following the encouragement of the ministry responsible for the opening of new research units, and in view of the explosion in the number of members of the laboratory which have exceeded 100 members between research teachers and students enrolled in Masters and doctoral thesis and in accordance with the recommendations during the last 571 evaluation of the 2003, several research units have been active members and responsible for research teams within the laboratory.

3. Calculating electrical requirements

To carry out the dimensioning of our grid-connected solar photovoltaic installation in the LASEM Laboratory at the National Engineering School, we propose two solutions for integrating the photovoltaic system. The first solution is to install the panels on the roof of the laboratory, facing south and at an angle of 30°. The second solution is to install the solar panels on the facades, facing south and at an angle of 90°.

The first step is to study the energy consumption of the laboratory. This step is very important if the system is to be optimal.

Kilowatt hours (kWh) are used to calculate annual electricity consumption.

All we need to do is multiply the appliance power P_a by the number of hours of use H_{uu}, then by the number of days J_u. Since the result is in Wh, we divide it by 1000 to obtain kWh, as shown in the following equation:

$$CAE\ (kWh) = \frac{((H_{uu}\ J_u\ P_a))}{1000} \quad (3.1)$$

Where :

H_{uu}: designates the unit hours of use per day

J_u: designates the days of use per year

P_u: denotes the unit power in watts (W)

P_a: designates the total power of the appliance in watts (W)

Nr: refers to the number of devices

Table 3.1 shows all the electrical requirements for the LASEM laboratory.

Table 3.2. Electrical requirements of the laboratory

Equipment	Nr	H_{uu}(h)	J_u (days/year)	P_u (W)	P_a (W)	CAE (kWh)
Computers	4	8	267	250	1000	2136
Tubes	6	12	267	40	240	768.96
air conditioners	1	10	79	3500	3500	2765
Video projector	1	5	40	624	624	124.8
Printer	2	1	267	650	650	173.55
Total	13	36	920	5064	6014	5968.31

According to the calculation of the electrical requirements of the LASEM laboratory summarised in Table 3.2, the electricity consumption in kWh is equal to 5968.31 kWh.

4. Sizing a grid-connected photovoltaic system on the roof

The first solution is defined by a grid-connected photovoltaic system integrated on the roof of the LASEM laboratory in Sfax. In order to size our PV installation, we need to identify the peak power.

4.1 The peak power of the photovoltaic system

The rated power of a PV module is called the peak power. It is expressed in W_p (Watt-peak) or kW_p (kilowatt-peak). In fact, the energy consumed per year is equal to the kWh/year and STEG tolerates a specific production of around 1kWp for every 1800 kWh/year of the adapted site.

The peak power of the installation Pc in kWp is calculated by the following equation

$$P_c = \frac{P_{com}}{Psp} \quad (3.2)$$

where

PC: Peak power of the installation in kWp

Pcom: Annual consumption power = 5968.31 kWh/year

Psp: Specific annual production for 1 kWp in kWh/year = 1800 kWh/year

According to the calculations, the peak power of the installation under consideration is equal to Pc = 3.315 kWp. In this application, to avoid the risks of overvoltage and circuit protection, we will increase the value of the peak power PC to 4 kW_c.

4.2 Components of the photovoltaic system

4.2.1 Photovoltaic panel

The solar module used in our study is a Canadian Solar Inc. module with a peak power PC = 410 Wp, of the class A monocrystalline type. It has an open-circuit voltage of Voc = 53.6 V and a short-circuit current of I_{sc}=9.70 A. The ABB 4.0 k delivers a voltage of V_{mpp}= 44.6 V and a current of $I_{(mpp)}$ = 9.08 A at the point of maximum power. This module complies with the NF IEC 61730 series of standards.

Figure 3.1. Canadian Solar Inc. photovoltaic panel.

Table 3.3 describes the necessary characteristics of the Canadian Solar Inc. mono-crystalline silicon solar panel that we chose for our installation. Other parameters can also be found in the data sheet attached in Appendix 1.

Table 3.3. Technical characteristics of the Canadian Solar Inc. panel CS1U - 410 MS

Module reference	CS1U - 410MS	
Maximum power (P_{max})	P_{mpp}	410 WP
Voltage at P_{max}	Vmp	44.5 V
Open circuit voltage	Voc	53.6 V
Current at P_{max}	Imp	9.23 A
Short circuit current	Isc	9.70 A
Panel length	L	2.078 m
Panel width	L	0.992 m

4.2.2 Photovoltaic inverter

This component plays a very important role in the installation. It transforms the direct current, which varies according to the luminosity, into alternating current, which can be fed into the STEG grid. This current can be injected into the STEG grid.The choice of our inverter is ABB and model UNO-DM-4.0-TL-PLUS, This new addition to the ABB family of single-phase inverters represents the optimum solution for residential installations. The brand offers a range of products, systems and services that enable customers to increase power capacity, improve grid reliability, increase energy efficiency and reduce environmental impact. In this way, ABB continues to shape the grid of the future.

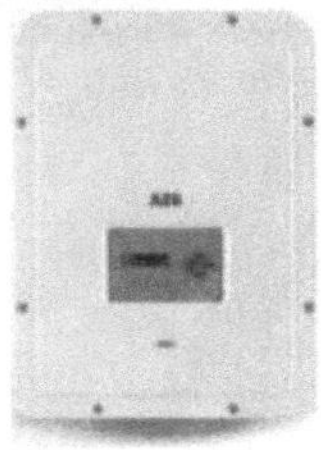

Figure 3.2. ABB 4 kW grid inverter

Table 3.4 describes the overall characteristics of the inverters used, as well as the electrical parameters of the inverter. In appendix 13, we can refer to the rest of its characteristics presented in its data sheet.

Table 3.4. Technical characteristics of the ABB inverter

Electrical parameters	**ue of the inverter**
Inverter reference	UNO-DM-4.0-TL-PLUS
DC	
Max DC power	3000 W
Max DC voltage	600 V
DC start input voltage Vstart	200 V (adj. 120...350 V)
MPP voltage range	0.7 x Vstart...580 V (min 90 V)
Max input current per MPPT	16 A /32 A
Number of MPPTs	2
AC	
Output power	4000 W
Max output power	4000 W
Max output current	17.2 A
Voltage range	180...264 V
Operating frequency	50/60 Hz
Power factor	> 0.995, ± 0.1 - 1
Mains connection	Single-phase
System	
Euro efficiency	96.5%
Degree of protection	IP 65

4.2.3 Installation configuration

To determine the number of Ppv panels required for this installation, calculate the ratio of total peak power to maximum panel power using the following equation:

$$N_{pv} = \frac{P_c}{P_{pv}} \quad (3.3)$$

With :

P_{pv}: Maximum power of a panel

P_c: Peak power.

According to our calculation, Npv = 9.75 panels. If we increase this value, we choose 10 panels instead.
In our installation, we have chosen to use an ABB inverter and 10 Canadian Solar Inc. model CS1U - 410 MS photovoltaic panels.
The minimum number (,,) $N_{(min)\ (>s)}$" and the maximum number (,,) $N_{(max)\ (>s)}$" of photovoltaic modules in series compatible with the MPPT voltage range of the inverter are expressed by the following equations:

$$N_{max,s} = PE_I\left(\frac{U_{mppt,max}}{U_{oc(-10°c)}}\right) \quad (3.4)$$

$$N_{min,s} = PE_S\left(\frac{U_{mppt,min}}{U_{mpp(80°c)}}\right) \quad (3.5)$$

With :
PEI: Lower integer part.
PES: Upper integer part.
U_{mp}pt,min: Minimum value of the tracker voltage (V).
U_{mp}pt,max: Maximum value of the tracker voltage (V).
The value of the open-circuit voltage of the photovoltaic modules $U_{(OC)\ (\_(10)\ "C)}$ (V) at a minimum temperature of Tmin = -10°C is calculated by the equation :

$$U_{oc(-10°c)} = K_u\ U_{oc(STC)} \quad (3.6)$$

Where Ku is the correction factor taking into account the increase in the open circuit voltage of the modules.
Considering the minimum ambient temperature Tmin = -10°C of the PV installation site and the temperature coefficient of the open-circuit voltage equal to a_{Uoc}= - 0.29 %, the correlation factor Ku is calculated by the following formula:

$$Ku = 1 + \left(\frac{\alpha_{Uoc}}{100}\right)(T_{min} - 25) \quad (3.7)$$

In our application, we find a correlation factor Ku = 1.1015
We can therefore deduce that the value of the open-circuit voltage of the photovoltaic modules is equal to $U_{(oc()\ \_(10)\ oc)}$= 59.04 V
Similarly, we can calculate the value of the MPP point voltage of the photovoltaic modules $U_{(mp)\ p}(80^{(o)\ c})$ *(Y)* at a maximum temperature equal to T_{max}= 80°C. It is calculated by the following equation:

$$U_{mpp(80°C)} = Ku\ U_{mp} \quad (3.8)$$

Under these conditions, the correction factor $K_u$takes into account the reduction in the closed-circuit voltage of the modules. Considering the maximum ambient temperature $T_{max}$of the photovoltaic installation site and the maximum power temperature coefficient equal to ttump $^{(=\_)\ 0}$.37%, the correlation factor $K_u$is calculated using the following formula:

$$Ku = 1 + \left(\frac{\alpha_{Ump}}{100}\right)(T_{max} - 25) \quad (3.9)$$

Under these calculation conditions, the correlation factor is Ku = 0.79.
Furthermore, the maximum power voltage of the photovoltaic modules at a temperature equal to T_{max}= 80°C is calculated by the following formula:

$$U_{mpp(80°C)} = Ku\ U_{mp} \quad (3.10)$$

Under these conditions, the maximum power voltage of the photovoltaic modules is equal to $U_{(mpp(80)\ oC)}$= 35.12 V

According to equations 3.4 and 3.5, this is :

$$N_{maxs} = 10 \text{ panels}$$

$$N_{min;s} = 3 \text{ panels}$$

The number of strings of photovoltaic modules in parallel compatible with the DC input current is expressed by the equation we find :

$$N_{max,p} = PE_I\left(\frac{I_{max}}{I_{mpp}}\right)$$

(3.11)

Therefore, $N_{max,p} = \frac{16}{9.23} = 1$ string

With :

PEi: Lower integer part,

$N_{max;P}$: Maximum number of strings in parallel,

I_{max}: Maximum permissible inverter MPPT current (A),

I_{mpp}: Maximum module power current (A).

In this installation, we have chosen an ABB inverter with 2 MPPT, each MPPT can accept a single string.

The number of panels per string $Nb_{(p)/c}h$ is calculated by the following formula:

$$N_{bp/ch} = N_{bp}\ N_{bch} \qquad (3.12)$$

With :

Nbp/ch : Number of panels per string

Nbp: Total number of panels = 10 panels

Nbch: Total number of strings = 2 strings

We have therefore chosen to install 2 strings in parallel. Each string contains Nbp/ch equal to 5 panels in series.

4.2.4 Inverter and PV array compatibility

4.2.4.1 Voltage compatibility

The maximum voltage of the modules in a photovoltaic string is calculated under winter conditions at a temperature of -10 °C. It must be lower than the maximum input voltage of the inverter.

The minimum module voltage in a photovoltaic string is calculated for a module temperature of up to 80° C. It must be higher than the minimum value of the inverter's MPPT voltage range.

Voltage compatibility for lows at -10°C enables the following inequality condition to be verified:

$$5\, U_{oc(-10°C)} < U_{max\ de\ l'onduleur} \qquad (3.13)$$

This condition is verified knowing that: 5 * 59.04 = 295.2 V < 600 V

Voltage compatibility for high temperatures at 80°C enables the following inequality condition to be verified.

$$5\, U_{mpp(80°C)} > U_{mpp,min} \qquad (3.14)$$

This condition is verified knowing that: 5 * 35.12 = 175.6 V > 90V

4.2.4.2 Current compatibility

The value of the DC current at the inverter input is :

- Strongly affected by solar radiation (direct linear relationship) ;

• Slightly affected by temperature (the current increases slightly as the temperature rises). The maximum current delivered by the photovoltaic generator must not exceed the maximum input current of the inverter $I_{(eo)(max)}$; i.e. :

$$I_{pv\ max} \leq I_{eo\ max} \tag{3.15}$$

The maximum current of the photovoltaic generator is determined from the number of strings connected in parallel according to this equation:

$$I_{sc\ (80°C)} = \left((T_{max} - 25) \times \frac{\alpha}{100} \times N_{max.p}\right) + I_{sc} \tag{3.16}$$

Thus, $I_{(sc(80)\ oC)}$= 9.755 A.
With :
T_{max}: Maximum temperature of the module taken equal to 80°C *a*: current/temperature coefficient of the photovoltaic module given by the module manufacturer in mA/°C is equal to 0.05 % / °C.
$N_{(max).p}$: Number of strings of photovoltaic modules in parallel equal to 2 strings.
I_{sc}: Open circuit current of the photovoltaic modules is equal to I_{sc}= 9.70 A.
maximum current= number of strings in parallellexl$_{mpp}$xl.25
maximum current = 2 x 9.23 (x)1.25 = 23.075 A < 32 A.

4.2.4.3 Power compatibility

The powers of the photovoltaic generator and the inverter must be mutually matched. The nominal AC power of the inverters can be equal to the power of the photovoltaic generator (in STC conditions) with a tolerance of ±10%.

$$0.8 \times P_n < P_{AC\,no} < 1.2 \times P_n \tag{3.17}$$

With :
P_n: Nominal power of the photovoltaic generator
PAC no: Nominal AC power of the inverter.
This inequation is verified knowing that :
0.8* 4000 = 3200 W <P_{ACnomi}naledel onduleur= 4000W < 1.2*4000 = 4800 W
The photovoltaic panels in the installation are therefore compatible with the chosen inverter.

4.3 Installing electrical safety tools

4.3.1 Connectors

MC4 connectors are quick to install, waterproof and very high performance. They make it easy to connect solar panels and your solar inverter. They are easy to assemble simply by inserting the connectors into each other, but they are secure because they require a tool to separate them. The choice of connectors depends on the following criteria:
The current-carrying capacity of the connector must be greater than :

$$Icc \times 1.25 = 9.70 \times 1.25 = 12.125\ A \tag{3.18}$$

In our installation, we used 4 mm^2MC4 connectors of type PV- KBT4/6I-UR and PV-KST4/6I-UR with a rated current of 39 A, a rated voltage of 600 V DC, protection class II, and a connected environmental protection degree of IP65 and IP68 (see appendix 14).

Figure 3.4. MC4 connector

4.3.2 DC enclosure

All the DC protection devices (DC surge arrester, switch-disconnector and fuse) are often grouped together in the same cabinet located at the inverter input (DC cabinet).

This installation will use two identical Mistral DC IDE enclosures for the 10 modules, with IP 65 protection, made by ZJ BENY.

4.3.2.1 Selecting the isolating switch

The choice of disconnector must take into account the following criteria:

- Rated operating voltage DC (U_e) greater than the maximum short-circuit voltage of the panels per string ($U_{comax/ch}$) i.e. :

$$Ue > Ucomax/ch \tag{3.19}$$

- Rated operating current $I_e$greater than the maximum service current I_{ms}i.e. :

$$I_e > I_{ms} \tag{3.20}$$

In the chain 1 disconnector, we have :

$U_{comax/ch}$= 295.2 V

I_{ms} = 1.25 Icc = 1.25X 9.70=12.125 A.

In this installation, the 2 strings are identical. We therefore choose the disconnector with the same maximum short-circuit voltage and the same service current.

The appropriate choice, for the two strings of this installation, two identical disconnectors of the

ZJ BENY with the following characteristics (see appendix 5):

- Rated operating voltage Ue = 600 VDC
- Rated operating current I_e= 16 A.

Figure 3.5. DC switch-disconnector

4.3.2.2 Choice of DC surge arrester

DC surge arrester chain 1 :

The choice of the DC surge arrester is based on the following criteria:

- $U_{cpv} > U_{oc\ max}$, with $U_{oc\ MAX} = 295.2$ V
- $U_{cpv} > U_{oc\ STCmax}$
- Up: Lightning arrester protection level, this value must be less than or equal to 80% of the surge withstand voltage U_{imp} of the electrical equipment to be protected
- ($Up < 0.8\ U_{imp}$)
- Imax = 40 kA
- In > 5 kA

Checking :

- $U_{cpv} > U_{oc\ MAX}$ with U_{oc} MAX at -10°C/chaÎne = 295.2 V
- Ucpv(v) > $U_{oc\ STC\ max}$

with :

Uoc STC max = 1.2 x Uoc STC = 321.6 V

(3.21)

$U_{oc\ STC} = Np\ '\ V_{oc} = 268$ V

- Up < 0.8 X Uimp

Up < 0.8 X 6.480 kV = 5.184 kV

- Imax = 40 kA
- In > 5kA.

The choice of the DC arrester for the two remaining chains is based on the same characteristics.

On the basis of the above criteria, four ZJ BENY type 2 DC surge arresters (see appendix 15) have been selected for this installation, with the following characteristics:

- U_{cpv}= 600 V: maximum steady state voltage.
- I_n= 20 kA: nominal discharge current.
- U_p < 3.8 kV: protection level.
- I_{max}= 40 kA: maximum discharge current.

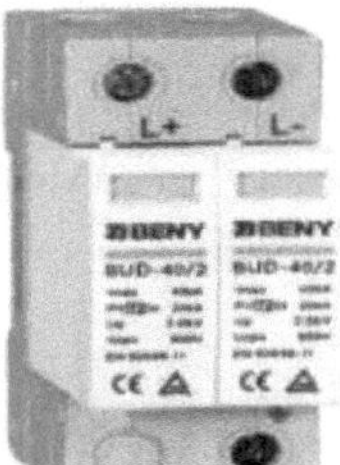

Figure 3.6. DC surge arrester

4.3.3 Choice of DC cables

It is recommended to use cables with a cross-section of 4 mm² in PV installations. We will therefore check that this cross-section is sufficient for our installation by means of two conditions:

- The admissible current.
- The maximum permissible voltage drop.

4.3.3.1 Voltage drop

This voltage drop is calculated under the following conditions:

- The resistivity of the cable is that corresponding to the maximum temperature of the core in normal service.
- The reference voltage used in the calculations is the STC voltage $U_{(mpp)}$.
- The reference current for the calculations is I_{mpp}STC.
- The voltage drop must be calculated for each PV string cable, each PV group cable, and for the main PV cable. The cumulative voltage drops of the cables between each string and the inverter are then calculated, and the highest cumulative value is retained.

For string 1 :

The voltage drop across the DC cable is written as :

$$\Delta U(V) = b\ R\ I$$

(3.22)

$$\Delta U(\%) = \frac{\Delta U \times 100}{U}$$

(3.23)

With: b = 2 (in continuous operation a double loss phenomenon).

I: nominal module current equal to = 9.23 A.

The resistance R is calculated from this formula:

$$R = \frac{\rho \times L}{S} = 0.225\ \Omega$$

Where :

L: length between photovoltaic field and inverter [m] equal to 40 m

p: copper conductivity at 90°C equal to 0.0225 Qmm²/m

S : Conductor cross-section equal to = 4 mm^2.

U : Nominal voltage of the module ᵡnumber of modules = 44.5ᵡ5= 222.5 V

$$\Delta U(V) = 4.15\ V < 6.9\ V$$

$$\Delta U(\%) = 1.8\ \% < 3\%$$

The same calculation of the voltage drop across the DC cable is repeated for the second strings, since both are identical.

The 4 mm^2cross-section of the DC cable is therefore sufficient for our installation according to the voltage drop conduction. (See Appendix 16).

4.3.3.2 Current carrying capacity

The ampacity of a cable is the maximum value of the current that can permanently flow through the conductor without its temperature exceeding its specified temperature.

Table 3.5. Permissible current IZ of photovoltaic cables for a maximum core temperature of 90°C (extract from UTE guide 32-502)

n 2 2 2 2\ Section (mm)	Two adjacent cables on wall
2,5	19 A
4	26 A
6	33
10	46 A
16	62 A

25	82 A
35	102 A

We have the admissible current $IZ = 1.25 \times I_{dc} = 1.25 \times 9.70 = 12.125$ A

Thus, according to Table 3.5, the 4 mm^2cross-section of the DC cable is sufficient for our installation.

4.3.3 AC enclosure

All the AC protection devices (circuit breaker, surge arrester) are generally in the same cabinet located at the inverter output (AC cabinet).

In this installation, we will be using a single enclosure with 10 watertight modules with IP 65 protection, made by ZJ Benny.

4.3.3.1 Choice of circuit breaker

An earth leakage circuit breaker (ELCB) is added to the inverter output circuit to isolate the inverter, provide an emergency shutdown and isolate the photovoltaic installation from the mains, provided it is easily accessible. The size of the connection circuit breaker is determined by the maximum current supplied by the inverter(s).

The circuit breaker in this installation is selected according to the following criteria:

- Rated voltage Uc = 230/400 V
- In > Is

With :

In: Circuit-breaker rated current.

Is: maximum inverter output current = 17.2 A.

We therefore use a single CHINT earth leakage circuit breaker with a rating of 25A and sensitivity of 30 mA and type NB1L-40 2P C25 30mA AC 6 kA (see Appendix 17).

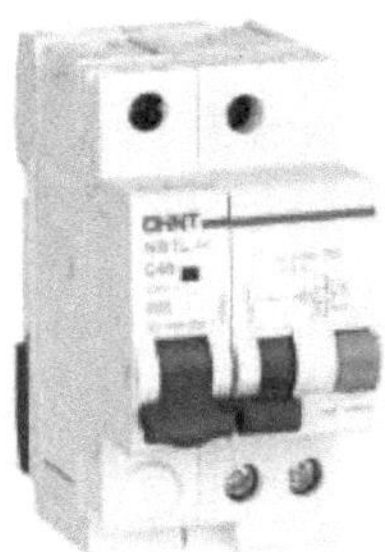

Figure 3.7. Earth leakage circuit breaker

4.3.3.2 Choice of AC surge arrester

The choice of AC surge arrester is based on the following criteria:

$I_n >> 5$ kA

$U_p << U_{surge} = 4000$ V (in accordance with guide UTE C 15-712-1)

$I_{max} >> 18$ kA.

In our installation, we must use two type 2 surge protectors made by ZJ Benny and type AC SPD 385V 2P BUA-40 IEC (see appendix 18), with the following characteristics:

- Us = 275 V
- In = 20 kA
- $U_p < 1.8$ kV
- $I_{max} = 40$ kA.

With :

- Us : Operating voltage
- In : Rated discharge current
- U_p: Protection level
- I_{max}: Maximum discharge current

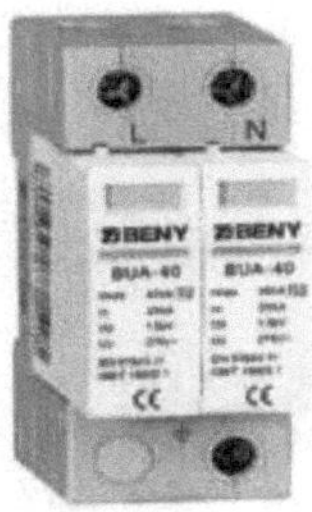

Figure 3.8. AC surge arrester

4.3.4 Choice of AC cables

For PV photovoltaic installations directly connected to the LV public distribution network, the maximum voltage drop permitted between the AC terminals of the inverter and the connection circuit breaker is 3% at the rated power of the inverter(s).

It is recommended to use cables with a cross-section of 4 mm^{2}in their PV installations. So we're going to check that this cross-section is sufficient for our installation.

The voltage drop across the AC cable at 90°C is calculated using the following formula:

AU = b(- X COS0 + ÀXLX SIN0) X I

(3.24)

With

- L = single conductor length [m] to STEG meter =3 m
- I = current per phase = 17.2 A
- P = copper conductivity at 90°C = 0.0225 mm2/m
- S = conductor cross-section [mm^2] = 4 mm^2
- U = single-phase voltage = 230 V
- b = 1 (for three-phase operation)
- SIN0 = 0
- COS0 = 1
- λ = 1 (cable reactance)

So we find AU= 0.29 V<< 6.9 V

AU(%) = (%) = 0.12% << 3 %

The 4 mm2 cross-section of the AC cable is therefore sufficient for our installation. (See Appendix 19)

4.4 Overvoltage protection description

To minimise the effects of induced overvoltages, the metal structures of the modules and the metal support structures (including the metal cable trays) must be connected to a protective equipotential bonding system which is itself connected to earth. As these metal structures are generally made of aluminium, suitable connection devices should be used. Copper conductors must not run in contact with aluminium parts. The earthing of PV modules is carried out in accordance with the manufacturer's instructions.

We have chosen the cross-section of the earthing cable for our installation from table 3.6 :

Table 3.6. Choice of protective conductor cross-section

Phase conductor section	Protective conductor section
Spb S Hi null'	SpE - Spj,
!6mm < Spit <35 nun'	Spi= 16 1ППГ
Ssp> 35 mm'	

In our installation the cross-section of all the phase conductors used is 4 mm^2done according to the table the cross-section of the earth cable must be 4 mm2, but according to the UTE C 15712-1 guide we will use a cable cross-section of 6 mm²as this is the minimum recommended cross-section.

4.4.1 Earthing the grounds and conductive elements of the DC section

Each structure is earthed via an earth cable with a cross-section of 10 mm2 using earth splitters. Earthing between the panels is also carried out using earth cables with a cross-section of 6 mm2.

4.4.2 Earthing the AC part of the installation

All the grounds on the AC side must be earthed by a protective conductor. The inverter is earthed via an earth cable with a cross-section of 10 mm2.

4.5 Structure of the photovoltaic installation against the wind

Our installation will use 10 panels in series. These panels are laid on slabs through triangles and protected against the wind by windbreaks.

In fact, the laying of the slabs is the basic pillar for the correct fixing of the photovoltaic field. Every 1.5 m, two panels are placed in parallel. The width of each panel is 0.992 m, so the total length of all the chains is 9.92 m.

So the total number of panels N

$$N = \frac{\text{La longueurs totale des chaînes}}{1.5} \times 2 = 13.22 \text{ dalles} \quad (3.25)$$

So in our installation we need to use 14 panels.

On each two parallel slabs, we fix a triangle. So we have 7 triangles.

The aluminium metal structure consists of 7 triangles made of 30^30x3 aluminium, each fixed to the roof by 2 concrete slabs (each slab weighs 65 kg). The triangles are reinforced with wind bracing.

These windbreaks are made of 30x30x3 aluminium, and are chosen so that the structure can withstand winds of up to 120 km/h.

The installation requires a photovoltaic array consisting of a 4.0 kWp system. This system consists of 10 monocrystalline modules of 410 $W_c$for a total of 4000 W_c. All the modules will be connected to an ABB 4.0 kWp inverter to supply AC voltage.

The system also includes protection boxes (two DC protection boxes and one AC protection box) to guarantee electrical protection on the photovoltaic generator side on the one hand and on the grid side on the other, in accordance with the standards for photovoltaic installations and the specifications.

The installation is made up of the components listed in table (3.7), and the technical data sheets for each component can be found in the corresponding appendix at the end of this report.

Table 3.7. Main components of the 4 kWp system

Description	Number	Brand	Reference	Appendix
Modules	10	Canadian Solar Inc.	HiDM CS1U-410 MS	Appendix 1
Inverter	1	ABB	ABB 4.0 KWC UNO-DM-4.0-TL-PLUS	Annex 13
Isolating switch DC	2	ZJ Benny	DC-PV1/DC-32B	Annex 5
Surge arrester DC	4	ZJ Benny	DC SPD TYPE 2 600V BUD-40/2 IEC	Annex 15
Surge arrester AC	2	ZJ Benny	ZJ BENY AC SPD - BUA-40/2	Annex 18
Earth leakage circuit breaker	1	Chint	NB1L-40 2P C25 30 mA AC6 KA	Appendix 17
DC cable	LOT	KBE SOLAR	Section 4 mm^2	Appendix 16
AC cable	LOT	Tunisia cable	Section 4 mm2	Appendix 19
Connector	LOT	Staubli Electrical	MC4 CONNECTOR	Appendix 14

5. Dimensioning of a photovoltaic system connected to the facades

For the renovation in our laboratory, we worked on integrating the photovoltaic system on the facades as the second solution with a 90° inclination and a southern orientation.

This type of system allows the photovoltaic panels to be used as a sunshade or to be integrated into the building. The panels are fixed to a supporting structure firmly anchored to the building structure. This type of installation requires planning permission.

In facades the energy needs are already calculated equal to = 5968.31 kWh/year but according to the technical reference of STEG the table gives a percentage variation between installation on 30° and 90°.

5.1 Calculating peak power

The power of the collector will decrease as the horizon decreases. It is reduced by 1/3 for a 90° inclination compared to a 30° inclination. The peak power is determined using the rule of three:

4^100% = 5.88 kWp

68%

In this application, we will increase the value of the peak power PC to 6 kWp.

5.2 Components of the photovoltaic system

5.2.1 Photovoltaic panel

The solar module used in our study is from Canadian Solar Inc. and has a power output of PC = 385 W_c. It is characterised by an open-circuit voltage V_{oc}= 48.0 V and a short-circuit current equal to I_{sc}= 10.09 A. The ABB UNO-DM-6.0-PLUS delivers a voltage V_{mpp}= 40.2 V and a current $I_{(mpp)}$ = 9.58 A at the point of maximum power. This module complies with the NF EN 61730 series of standards.

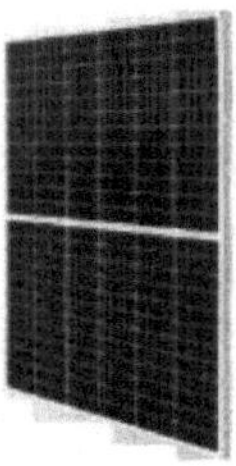

Figure 3.9. Photovoltaic panel at 385 wc

Table 3.8 describes the necessary characteristics of the Canadian Solar Inc. mono-crystalline silicon solar panel that we chose for our façade-integrated installation.

Other parameters are also shown in the data sheet attached in Appendix 20.

Table 3.8. Technical characteristics of the CS3U-385 Wp panel

Module reference	CS3U-385 MS 1500 V	
Maximum power (Pmax)	Pmpp	385
Voltage at Pmax	Vmp	40.2
Open circuit voltage	Voc	48
Current at Pmax	Imp	9.58
Short-circuit current	Isc	10.09
Panel length	L	2
Panel width	L	0.992

5.2.2 Photovoltaic inverter

Our choice of inverter is an ABB model UNO-DM-6.0-TL-PLUS, which is an optimum solution for large single-phase residential installations.

Figure 3.10. ABB 6 kWp solar inverter

Table 3.9 describes the overall characteristics of the inverters used, as well as the inverter's electrical parameters. In appendix 21, we can refer to the rest of its characteristics presented in its technical data sheet.

Table 3.9. Technical characteristics of the ABB 6 kWp inverter

Electrical parameters of the inverter	
Inverter reference	UNO-DM-6.0-TL-PLUS
DC	
Max DC power	4000 W

Max DC voltage	600 V
DC start input voltage Vstart	200 V (adj. 120...350 V)
MPP voltage range	0.7 x Vstart...580 V
Max input current	25 A
Number of MPPT	2
AC	
Output power	6000 W
Max output power	6000 W
Max output current	30 A
Voltage range	180...264 V
Operating frequency	50 Hz / 60 Hz
Power factor	> 0.995± 0.8 - 1
Mains connection	Single-phase

System	
Euro efficiency	97.0%
Degree of protection	IP 65

5.2.3 Installation configuration

To determine the number of panels required for this installation, the ratio of the total peak power to the maximum power of the panel is calculated using the following equation: Npv =

$$N_{pv} = \frac{P_c}{P_{pv}} \qquad (3.26)$$

where

P_{pv}: Maximum power of a panel

P_c: Peak power

According to our calculation, Npv = 15.75 panels. By increasing this value, we choose 16 panels instead.

In our installation, we chose to use an ABB inverter and 16 Canadian Solar Inc. CS3U-385 MS 1500V photovoltaic panels.

According to equations (3.4), (3.5) and (3.6) and considering the minimum ambient temperature Tmin = -10°C of the PV installation site and the temperature coefficient of the open circuit voltage equal to a_{Uoc}= -0.29 %, the correlation factor $K_u$is calculated by formula (3.7) and we find Ku = 1.1015.

We can therefore deduce that the open-circuit voltage of the photovoltaic modules is equal to $U_{(oc()\_(10)\ oc)}$= KuUoc=52.9 V.

Similarly, we can calculate the value of the MPP point voltage of the photovoltaic modules $U_{(mp)\ p}(80^{(o)\ c})$ *(V)* at a maximum temperature equal to T_{max}= 80°C.

Under these conditions, the correction factor Ku takes into account the decrease in the closed-circuit voltage of the modules. Considering the maximum ambient temperature $T_{max}$of the photovoltaic installation site and the maximum power temperature coefficient equal to ^{a} ump$^{=}$ -0.36 %, the correlation factor Ku is calculated using equation (3.9)

Under these calculation conditions, the correlation factor is equal to Ku = 0.80.

Furthermore, the maximum power voltage of the photovoltaic modules at a temperature equal

to T_{max}= 80°C is calculated by equation (3.8).
Under these conditions, the maximum power voltage of the photovoltaic modules is equal to $U_{(mpp)((80) oC)}$= 32.16 V
In our case this is

$$N_{max,s} = \frac{600}{52.9} = 11 \text{ panneaux}$$

$$N_{min,s} = \frac{90}{32.16} = 3 \text{ panneaux}$$

The number of strings of photovoltaic modules in parallel compatible with the current between DC is equal to $N_{max>p}$= 2 strings and the number of panels per string Nbp/$_c$h is equal to 8 panels.
We therefore chose to install two chains in parallel. Each string contains 8 panels in series.

5.2.4 Inverter and PV array compatibility

5.2.4.1 Voltage compatibility.

Voltage compatibility for lows at -10°C allows the following inequality condition to be verified:

$$8\, U_{oc(-10°C)} < U_{max\ de\ l'onduleur} \quad (3.27)$$

This condition is verified knowing that: 8 * 52.9 = 423.2V < 600 V
Voltage compatibility for high temperatures at 80°C enables the following inequality condition to be verified.

$$8\, U_{mpp(80°C)} > U_{mpp,min} \quad (3.28)$$

This condition is verified knowing that: 8* 32.16 = 257.28V > 90V

5.2.4.2 Current compatibility

The maximum current delivered by the photovoltaic generator must not exceed the maximum input current of the inverter (max current <=$I_e max$)■
The maximum current of the photovoltaic generator is determined from the number of strings connected in parallel according to equation (3.16) $I_{(sc)}(8o^{(o)\,c})$ = 10.145 A
maximum current= number of parallel strings * I_{mpp}* 1.25
maximum current = 2 * 9.58 * 1.25 = 23.95A < 40 A

5.2.4.3 Power compatibility

The rated AC power of the inverters can be equal to the power of the photovoltaic photovoltaic generator (under STC conditions) with a tolerance of ±10%.

$$0.8xP_n < P_{(ACno)} <1.2xP_n$$

In our case :

$$4800\ W < P_AC \text{ nominal inverter} = 6000W <= 7200\ W$$

Done, the photovoltaic panels in the installation are compatible with the chosen inverter.

5.3 Setting up the electrical safety tools for the installation

5.3.1 Connectors

In our installation we use the same MC4 connectors type PV-KBT4/6I-UR and PV-KST4/6I-UR (see appendix 14) with a rated current of 39 A.

5.3.2 DC enclosure

All the DC protection equipment (DC lightning arrester, switch-disconnector and fuse) are often grouped together in the same cabinet located at the inverter input.
In this installation we will be using two identical Mistral IDE DC enclosures, for the 16 modules, with IP 65 protection, made by ZJ BENY.

5.3.2.1 Choosing the isolating switch

The choice of disconnector must take into account the following criteria:

- $U_e > U_{comax/ch}$
- $I_e > I_{ms}$

Chain 1 disconnector

In the chain 1 disconnector, we have :

$U_{comax/ch}$= 423.2 V

I_{ms} = 1.25 Icc = 1.25x910.09 =12.6125 A

In this installation, the 2 chains are identical. We therefore choose the disconnector with the same maximum short-circuit voltage and the same service current.

The appropriate choice, for this installation, two identical disconnectors of the brand DIGITAL ELECTRIC with the following characteristics (see appendix 22):

- Rated operating voltage 600 VDC
- Rated operating current 25 A

Figure 3.11. DC disconnector

5.3.2.2 Choice of DC surge arrester

DC surge arrester chain 1 :

The choice of DC arrester is based on the following criteria:

- Ucpv > U_{ocMAx}, with UOCMAX = 423.2 V
- Ucpv UocSTCmax >
- Up: Lightning arrester protection level, this value must be less than or equal to 80% of the surge withstand voltage Uimp of the electrical equipment to be protected
- (Up<0.8 Uimp)
- Imax = 40KA
- In > 5kA

Checking :

- Ucpv > UOCMAX with UOCMAX at -10°C/sheath =423.2V
- Ucpv(v)> UocSTCmax

With UocSTCmax = 1.2 x Uoc STC=1.2x384 = 460.8V

Uoc STC = number of panels x Voc = 8 x48 = 384V

- Up < 0.8 x Uimp

Up < 0.8 x 6.480 kV = 5.184 kV

- Imax = 40 kA
- In > 5kA

DC surge arrester chain 2 :

The choice of the DC arrester is made using the same criteria as for the DC arrester in chain 1. Based on the above criteria, we have chosen four type 2 600 V DC surge protectors for our installation, made by DIGITAL ELECTRIC (see appendix 23), with the following characteristics:

- Ucpv = 600 V: maximum steady state voltage
- In = 20 Ka: nominal discharge current
- Up <3.6 kV: protection level
- Imax = 40 kA: maximum discharge current.

Figure 3.12. DC surge arrester

5.3.3 Choice of DC cables

It is recommended to use cables with a cross-section of 4 mm^{2}in PV installations. We will therefore check that this cross-section is sufficient for our installation by means of two conditions:

- The admissible current.
- The maximum permissible voltage drop.

5.3.3.1 Voltage drop

The voltage drop is calculated using the following formula: AU(V) = b X R X I

$$\Delta U(V) = b \times R \times I$$

$$\Delta U(\%) = \frac{\Delta U \times 100}{U}$$

Where: b = 2 (in continuous operation a double loss phenomenon).

I= nominal module current = 9.58 A.

$$R=\frac{\rho \times L}{S} = \frac{0.0225\times 40}{4} = 0.225\ \Omega$$

Or

L : length between photovoltaic field and inverter [m] = 40 m p = 0.0225 Qmm^2/m : conductivity of copper at a temperature of 90°C

S : Conductor cross-section in [mm^2] = 4 mm^2.

U: Nominal module voltage x number of modules= 40.2x8=321.6V.

$$\Delta U(V) = b \times R \times I\,[V]$$
$$= 2 \times 0.225 \times 9.58$$
$$\Delta U(V) = 4.31V < 6.9\ V$$
$$\Delta U(\%) = \frac{\Delta U \times 100}{U} = \frac{4.31 \times 100}{321.6} = 1.3\,\% < 3\%$$

The same voltage drop calculation is repeated for the second chain, since the two chains are identical.

The 4 mm2 cross-section of the DC cable is therefore sufficient for our installation according to the voltage drop calculation. (See Appendix 16)

5.3.3.2 Current-carrying capacity

The admissible current IZ of the photovoltaic cables is as follows

IZ = 1.25 X Iccl.25 X 10.09 = 12.61 A

Therefore, according to table (3.7), the 4 mm2 cross-section of the DC cable is sufficient for our installation.

5.3.4 AC enclosure

All the AC protection devices (circuit breaker, lightning arrester) are generally in the same cabinet located at the inverter output (AC cabinet). In this installation, we will be using a single enclosure with 10 watertight modules with IP 65 protection, made by ZJ Benny.

5.3.4.1 Choice of circuit-breaker

The choice of circuit breaker for this installation is based on the following criteria:

- Rated voltage Uc = 230/400 V
- In> Is

With :

- In: circuit-breaker rated current.
- Is: maximum inverter output current = 30 A.

The appropriate choice for this installation is a SCHNEIDER 40 A differential circuit breaker with 30 mA sensitivity, type Acti9 iDD40K (see appendix 24).

Figure 3.13. AC earth leakage circuit breaker

5.3.4.2 Choice of AC surge arrester

To protect the inverters against any surges from the distribution network, it is often necessary to install a lightning arrester with its own specific protection at the main LV board.

The choice of AC surge protector is based on the following criteria:

- In >> 5 kA
- Up <<U surge = 4000 V (according to guide UTE C 15-712-1)

- Imax >> 18 kA

A DIGITAL ELECTRIC type 2 AC surge protector must be used (see appendix 25), with the following characteristics:

- Us = 320 V
- In = 20 kA
- Up < 1.5 kV
- Imax = 40 kA

Figure 3.14. AC surge arrester

5.3.5 Choice of AC cables

They recommend using cables with a cross-section of 4 mm²in their PV installations. So we're going to check that this cross-section is sufficient for our installation.

Voltage drop across the AC cable at 90°C :

$$\Delta U = b(\frac{\rho \times L}{s} \times COSØ + \lambda \times L \times SINØ) \times I$$

Where :

- L = single conductor length [m] to STEG meter =3 m
- I = current per phase = 30A
- P = copper conductivity at 90°C = 0.0225 mm2/m
- S = conductor cross-section [mm^2] = 4 mm^2
- U = single-phase voltage = 230 V
- b = 1 (three-phase)
- SINØ = 0
- COSØ = 1
- λ = 1 (cable reactance)

$$\Delta U = \frac{0.0225 \times 3}{4} \times 30 = 0.50 \text{ V} << 6.9 \text{ V}$$

$$\Delta U(\%) = \frac{\Delta U \times 100}{U}(\%) = \frac{0.50 \times 100}{230}\% = 0.21\% << 3\ \%$$

The 4 mm2 cross-section of the AC cable is therefore sufficient for our installation. (See Appendix 19)

In conclusion, our installation requires a photovoltaic generator consisting of a 6.0 kWp system. This system is made up of 16 monocrystalline modules of 385 Wp for a total of 6000 Wp. All the modules will be associated with an inverter, ABB 6.0 K, in order to supply an AC voltage, and it also includes protection boxes (two DC protection boxes and a single AC protection box).

Our PV installation on the facades of the laboratory consists of the elements listed in Table (3.10), and the data sheets for each element are given in the relevant appendix at the end of this report.

Table 3.11. Main components of a 6 kWp installation

Description	Number	Brand	Reference	Appendix
Modules	16	Canadian Solar Inc.	SKU CS3U-385 MS 1500V	Appendix 20
Inverter	1	ABB	ABB 6.0 KWC UNO-DM-6.0-TL-PLUS	Annex 21
Isolating switch DC	2	DIGITAL ELECTRIC	isolating switch 25A 600V dc 04312	Annex 22
Surge arrester DC	4	DIGITAL ELECTRIC	DC SPD TYPE 2 600V 08402	Schedule 23
Surge arrester AC	2	DIGITAL ELECTRIC	AC Surge Arresters (Bi/Tri/Tetra) 08204	Annex 25
Earth leakage circuit breaker	1	Schneider	A9DK1640	Appendix 24
DC cable	LOT	KBE SOLAR	Section 4 mm^2	Appendix 16
AC cable	LOT	Tunisia cable	Section 4 mm2	Appendix 19
Connector	LOT	Staubli Electrical	MC4 CONNECTOR	Appendix 14

6. Comparison of the I-V electrical characteristics between the panel and the protection cabinets.

In the following table (3.12) we have compared the voltage and current of the panel used with the voltages and currents of the DC/AC enclosures chosen:

Table 3.11. I-V characteristics of the components of an IPV on the roof

		DC box		AC cabinet	
	Panel	Disconnector	Surge arrester	Surge arrester	Circuit breaker
Voltage (V)	53.6	600	600	275	230
Current (I)	9.70	16	40	40	25

The electrical I-V characteristics between the panel and the other components of a façade-integrated grid-connected photovoltaic system are compared in the following table:

Table 3.12. I-V characteristics of the components of a façade-mounted PV system

		DC box		AC box	
	Panel	Disconnector	Surge Protector	Surge arrester	Circuit breaker
Voltage (V)	48	600	600	320	230
Current (I)	10.09	25	40	40	40

7. Simulation of installations using PVsyst software

The PVsyst software was used to size two photovoltaic systems, one for a grid-connected photovoltaic installation on the roof and the other for an installation on the facades of the laboratory.

We simulated our photovoltaic system on the roof with a power output of
4.0 kWp and on the facades it becomes 6.0 kWp using the PVsyst version 6.8.1 software.

7.1 Grid-connected photovoltaic system 4 kWp

With the help of this software we have studied and analysed the project in detail for a grid-coupled system that defines the main input parameters such as orientation. For the 4.0 KWp photovoltaic installation the type of PV field is a fixed inclined plane with a 30° tilt and south orientation.

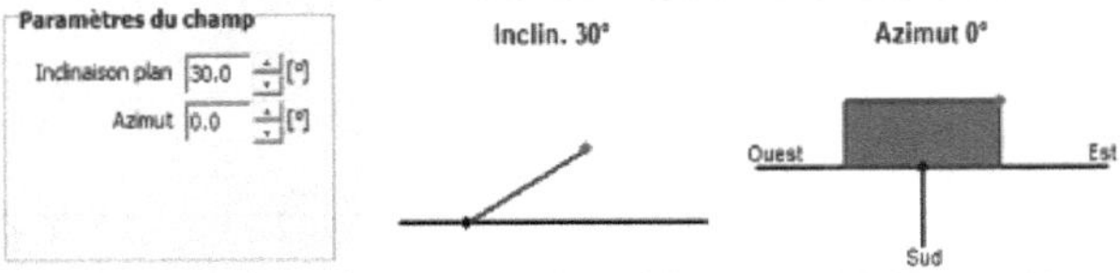

Figure 3.14. Orientation and inclination of panels on the roof

Our grid-connected rooftop photovoltaic installation requires configuring the choice of panel and inverter compatible with the 4 kWp peak power as shown in the following figure (see Appendix 26).

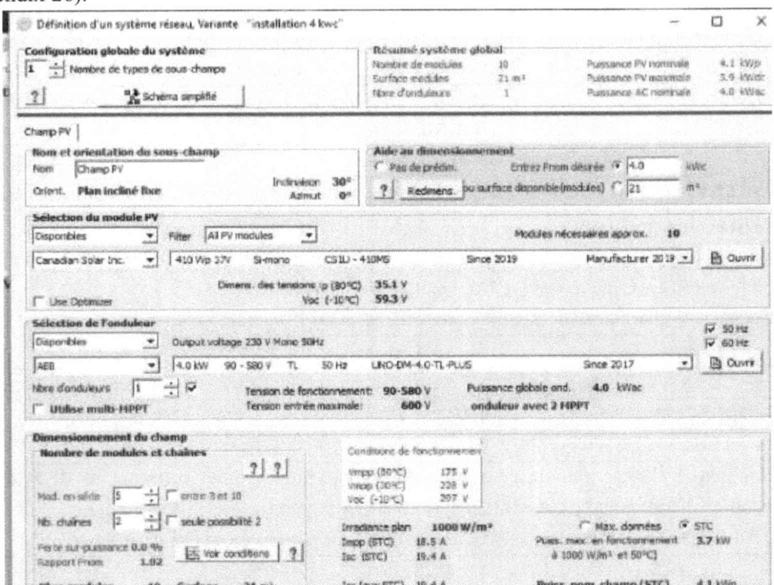

Figure 3.15. Configuration of the 4.0 kWp PV system

7.2 Grid-connected PV system 6 kWp

The difference between the 4 kWp grid-connected rooftop PV system and the 6 kWp facade PV system is the change in orientation of the PV array, which is shown in the following figure:

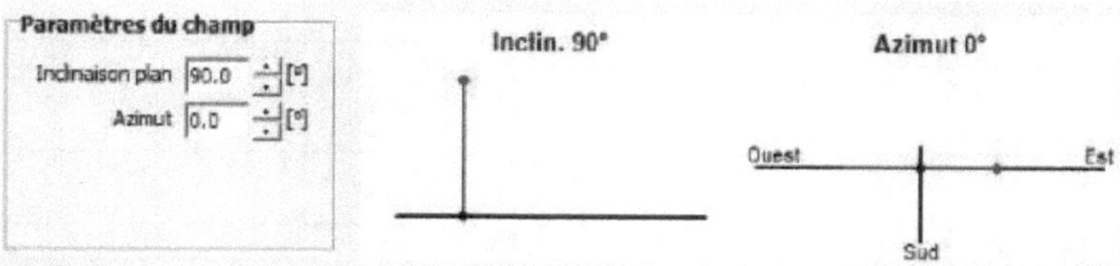

Figure 3.16. Orientation and tilt of panels on facades

The sizing of this PV installation is based on the choice of a panel and an inverter compatible with the 6.0 KWp peak power (see Appendix 27).

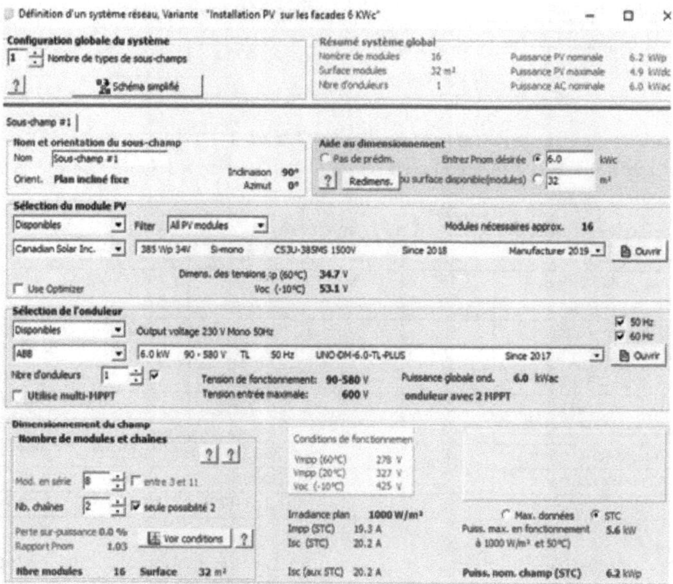

Figure 3.17. PV 6.0 kWp system configuration

8. Economic study of PV installations

To determine the cost of a photovoltaic installation, you need to know the price of the photovoltaic modules, the cost of the inverters, the cost of the electrical appliances and the cost of the photovoltaic cables. Then there's the cost of labour, which is an essential part of the quality of a photovoltaic installation.

8.1 Grid-connected photovoltaic installation on the roof

On the roof of the laboratory we have integrated a grid-connected photovoltaic system with an output of 4.0 kWp. Table (3.13) shows the prices of the various components of our photovoltaic installation:

With :

- Qty: Quantity of components used
- VAT%: value added tax
- PU: the unit price of the component
- PTHT: total price excluding tax
- VAT: all taxes included

Table 3.13. Quotation for a 4.0 kWp PV installation

Number	Description	Quantity	VAT	PU	PTHT	INCL. VAT(DT)
1	Photovoltaic module 410 WC	10	7%	750.000	7500.000	8 025 000
2	ABB inverter 4.0	1	7%	3700.000	3700.000	3959 000
3	DC protection box	2	19%	250.000	500.000	595 000
4	AC protection box	1	19%	220.000	220.000	261 800
5	Metal structure in aluminium	1	19%	1.000.000	1.000.000	1 190 000
6	Wiring and earthing	1	19%	200.000	200.000	238 000
7	Installation costs	1	19%	700.000	700.000	833 000
8	Stamp					0.600
Total				6 820 000	13 820 000	**13 912 400**

8.2 Grid-connected photovoltaic system on the facades

The cost of our 6.0 KWp grid-connected photovoltaic system integrated on the facades of the LAZEM Laboratory is shown in the following table:

Table 3.14. Cost estimate for a 6.0 kWp PV installation

Number	Description	Qty	VAT	PU	PTHT	INCL. VAT
1	Photovoltaic module 385 WC	16	7%	600.000	9600.000	10 272 000
2	ABB 6.0 inverter	1	7%	5100.000	5100.000	5 457 000
3	DC protection box	2	19%	270.000	540.000	642 600
4	AC protection box	1	19%	250.000	250.000	297 500
5	Aluminium metal structure	1	19%	1700.000	1700.000	2 023 000
6	Wiring and earthing	1	19%	350.000	350.000	416 500
7	Installation costs	1	19%	10000.000	1000.000	1 190 000
Total				9 270 000	18 540 000	**19 924 350**

According to tables (14) and (15), we have compared the percentage cost between a grid-connected PV installation integrated on the roof and a PV installation integrated on the facades in the laboratory is equal to 43.21%.

9. Conclusion

In this chapter, we have carried out a theoretical study of a small-scale grid-connected photovoltaic installation in two conditions, one on the roof and the other integrated on the facades for the LAZEM Laboratory. Firstly, we have calculated the energy requirements and then selected all the equipment required for the two installations. Secondly, we validated the dimensions of the installations using the PV syst photovoltaic software. Finally, we studied and compared the cost of the PV systems on the façades of the laboratory, which is 43.2% higher than on the roof. Therefore it is not a feasible solution for the vertical building.

Chapter 4

Experimental study of the energy parameters of a vertical building

1. Introduction

In this chapter, we are interested in the experimental study of climatic conditions for vertical buildings specifically within the Laboratory of Electromechanical Systems (LASEM) by specifying the main parameters of thermal comfort as well as the study of the characteristics of a photovoltaic panel. The measurements were carried out in June and July 2021 and were obtained using an acquisition system integrated into the closed chamber to ensure thermal comfort and an acquisition system integrated into the PV panel. The aim of this chapter is to compare the results obtained in the experiment for the two cases (with and without cooling of the PV panel) and for the experiment in the laboratory we compare the distribution of temperature and humidity for different modes.

2. Presentation of the equipment

In this chapter, the experimental study is divided into two different parts, the first to study thermal comfort in the laboratory and the other to ensure the energy comfort of a photovoltaic system.

2.1 Experimental set-up

2.1.1 First experiment

The experimental set-up is presented by a photovoltaic solar panel. The experimental study took place in the electromechanical systems laboratory (LASEM) in collaboration with the company SENCOM. The experimental study was carried out using an acquisition system based on an Arduino Uno board, two test plates, a Victron 75/15 charge regulator and a DC load, all of which were installed on the collector support to monitor its operation. This PV sensor is oriented to the south of Sfax whose geographical coordinates are Latitude is equal 34.725 N and Longitude is equal 10.717 E and tilted with a 30° degree.

Figure 4.1. Photovoltaic solar panel.

2.1.2 Second experiment

The experimental set-up is presented by DHT22 collectors in a chamber. The experimental study took place in the LASEM laboratory at the National Engineering School. This experimental study was carried out using an acquisition system based on an Arduino Uno board, a test plate and six humidity and temperature sensors connected to the computer by a USB cable to monitor its operation.

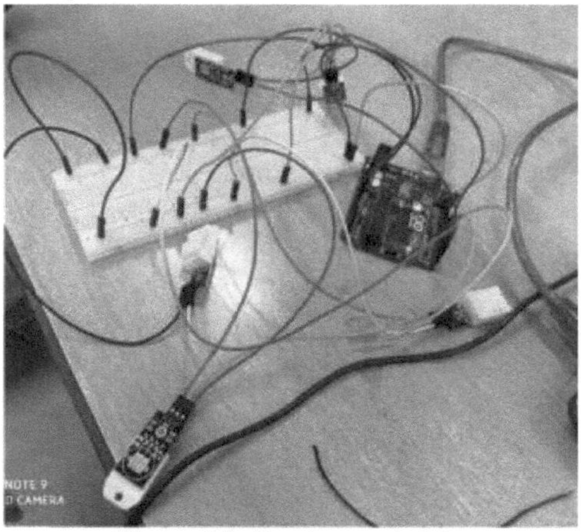

Figure 4.2: Structure materials.

2.2 Acquisition system components

2.2.1 Photovoltaic panels

There are several manufacturers of photovoltaic panels in the world. In this experimental study, we worked with the SCHOTT SOLAR POWER POLY 250W polycrystalline module, which functions as an electrical current generator in the presence of solar radiation. This thin-film module has a high energy yield in diffused light conditions, on sites exposed to heat and without rear ventilation, and on partially shaded roofs without optimum orientation. Solar modules with polytextured cells are made from high-quality crystalline solar cells to ensure maximum energy yield. They have received top marks in a number of independent studies and surveys. Figure 4.3 shows the dimensions of the panel, which has a length Lp = 1625 mm, a
width lp = 990 mm and thickness ep = 35 mm and the nameplate of one panel.

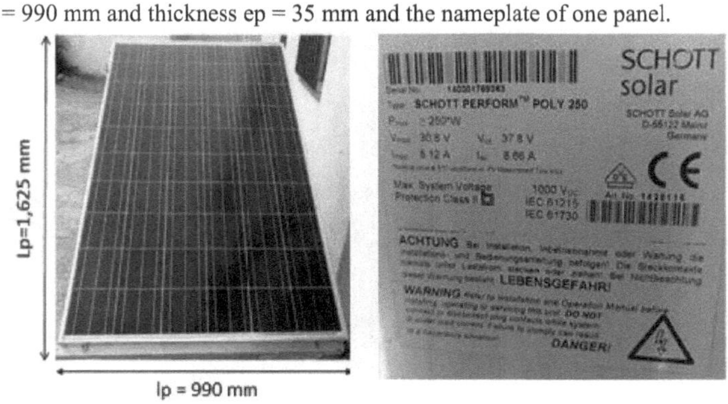

(a) Panel dimensions (b) Nameplate

Figure 4.3. Polycrystalline photovoltaic panel

The SCHOTT SOLAR POWER POLY solar module used in this study has a peak power equal to PC equal to 250 WC. It is characterised by an open-circuit voltage VOC = 37.8 V and a short-circuit current ISC = 8.66 A. With an efficiency of up to 15.3%, it delivers a voltage VMPP = 30.8 V and a current IMPP = 8.12 A at the point of maximum power. The module's electrical parameters are summarised in the following table (see Appendix 30).

Table 4.1. Technical characteristics of the PV in STC

Electrical parameters (Standard Test Conditions - STC)	
Model reference	SCHOTT SOLAR POWER POLY 250W
Maximum power (P_{max})	250 WP
Voltage at Maximum Power (V_{max})	30.8 V
Short circuit current (I_{sc})	8,66 A
Open Circuit Voltage (V_{oc})	37.8 V
Current at Maximum Power (C_{max})	8.12 A
Module Efficiency	15.3%

2.2.2 Sensor support

There are several methods for installing photovoltaic panels, including fixing the panels to a support. This support carries the envelope with the collector and holds them at a 30° angle to the ground. It also contains the electronic components. It is made of 40x40x4 mm aluminium angle and is occupied by 4 wheels for ease of transport. Figure 4.4 illustrates the different support components of the photovoltaic collector.

N°	**Designation**
1	PV collector
2	Aluminium corner pieces
3	Plexiglass plate
4	Wheel

Figure 4.4. Sensor support

2.2.3 Arduino board

Arduino is the brand name of an open-source prototyping platform that enables users to create interactive electronic objects from electronic boards. The Arduino UNO board, the most popular of the Arduino boards, is a small electronic board equipped with a microcontroller.

The microcontroller is used to program and control actuators based on events detected by sensors. All you have to do is connect it to a computer using a USB cable, with power supplied by the USB port. The Arduino board is therefore a programmable interface, as shown in Figure (4.5).

Figure 4.5. Arduino Uno board

The main characteristics of an Arduino Uno board are described in table 4.2:

Table 4.2. Characteristics of an Arduino board

Power supply	via 7-12 V USB port
Microprocessor	ATMega328
Flash memory	32 KB
SRAM memory	2 KB
EEPROM Memory	1 KB
Interfaces	14 I/O pins including 6 PWM
Current per I/O	40 mA
Timing	16 MHz

A description of all the connectors on the Arduino Uno board is shown in Figure 4.6 :

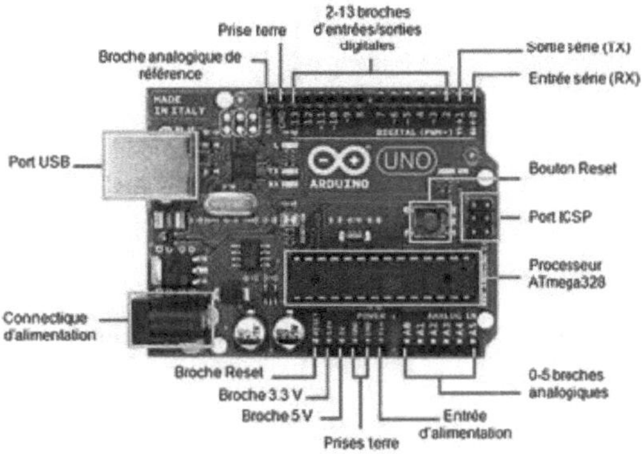

Figure 4.6. Description of the inputs/outputs on the Arduino Uno board

2.2.4 Charge controller

A solar charger recovers energy from your solar panels and stores it in your batteries. Using the latest and fastest technology, Blue Solar maximises this energy 'harvest', intelligently managing it so that you get a full charge as quickly as possible. Figure 4.7 shows the charge controller installed in the photovoltaic system. This Blue Solar MPPT 75/15 controller is connected to the photovoltaic sensor (see Appendix 31), and the technical characteristics of each controller are shown in Table 4.2.

Table 4.3. Characteristics of the Blue Solar MPPT 75/15

Designation	Specifications
Battery voltage	12/24 V
Nominal charging current	15 A
Maximum PV voltage	75 V
Operating temperature	-30 to +60 °C

Figure 4.7. Blue Solar MPPT 75/15 charge controllers

2.2.5 Charging lamps

In this experiment we have 4 12 V DC indicator lamps as the source of the load to measure the electrical characteristics is essentially the power of a photovoltaic panel photovoltaic panel using a victron Connect application installed on the laptop.

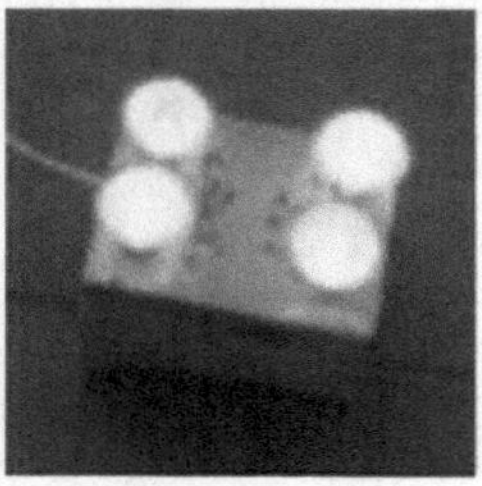

Figure 4.8. Load lamps.

2.2.6 Victron Connect application

Use the Victron Connect application to configure, monitor, update and diagnose your Victron product. Available on devices running Android, Windows, and MacOs, it can connect to Victron products via Bluetooth, USB, and Wi-Fi. In our photovoltaic panel experiment we connected the controller with the DC load and the Victron Connect application using the USB type C cable and USB connector. This application displays the power P, voltage U, and current I of the load.

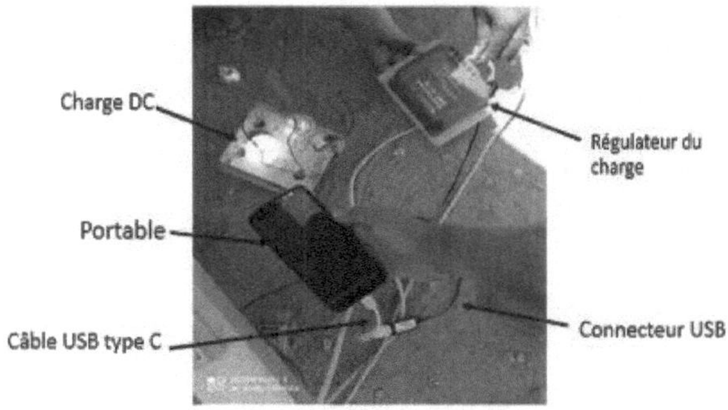

Figure 4.9. Application method

2.3 Measuring instruments

2.3.1 Thermocouple DS18B20

To study the performance of the PV sensor, it is necessary to measure the temperature of the device's components. To do this, we installed DS18B20 thermocouples at the input and output of each component. The DS18B20 digital thermometer provides temperature measurements in degrees Celsius with a value coded from 9 to 12 bits. Its measurement range extends from -55°C to +125°C with an accuracy of ± 0.5°C. It has an alarm function with high and low trigger points that are non-volatile and user-programmable. Each thermocouple has a unique 64-bit serial number, allowing multiple DS18B20s to operate on the same 1-wire bus. This makes it easy to use an Arduino Uno board to control many DS18B20s spread over several positions. To do this the position of the thermocouples mounted on the axis (ox) and the two axes (oz) of the board. Figure (4.10) shows the distribution of the thermocouples on the back of the PV sensor and Figure (4.11) shows a photo of a DS18B20 thermocouple. (See Appendix 32).

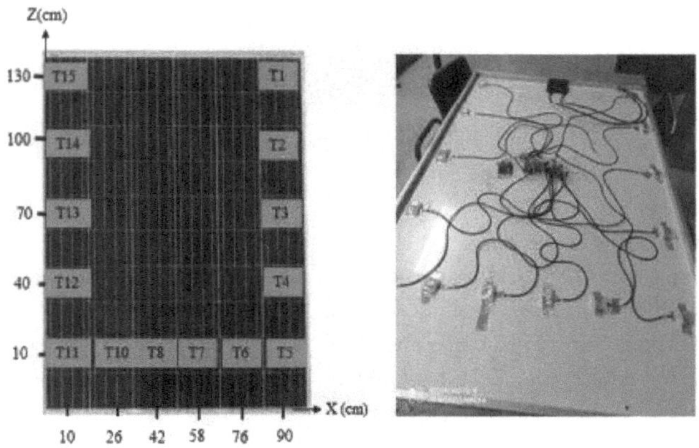

Figure 4.10. The distribution of thermocouples

Figure 4.11. DS18B20 temperature sensor

2.3.2 Solarimeter

The experimental study of the system requires a series of measurements to characterise the temperature and solar radiation on the slope of the panel. Parameters that are constant throughout the experiment, such as solar irradiance, are measured. For other parameters, such as temperature, which vary at each instant, we opted to use a data acquisition system to record the data. Figure (4.12) shows the solarimeter, an instrument used to measure and control solar power - cumulative energy in Wh/m^2- instantaneous, average, min/max values (see Appendix 33). All you have to do is take the SPM 72 and point its aperture towards the sun to measure the power of its radiation, which is fixed directly to the surface of the panel, enabling you to measure, in watts per square metre, the intensity of solar radiation received at a given point. Table (4.4) illustrates the technical characteristics of the solarimeter. The power received by the photovoltaic panel Pr can be calculated from the following formula:

$$Pr = A\,Gt \qquad (4.1)$$

Where :

A is the surface area of the panel and Gt is the global solar radiation measured by the solarimeter.

Figure 4.12. Solarimeter.

Table 4.4. Characteristics of the solarimeter.

Designation	Characteristics
Maximum working height	2000 m
Resolution	0.01 m/s
Operating temperature	18°C to 28°C below 75% RH
Sampling time	0,4 s
Length	165 mm long
Width	85 mm
Height	38 mm
Weight	200 g

2.3.3 DHT22 sensor

To monitor the humidity of the ambient environment as well as that of the closed chamber, we opted to use the DHT22 type sensor, capable of sending both temperature and humidity values at the same time. This sensor is easy to supply at low cost. It can be communicated serially using a special protocol, and provides a temperature indication on
16-bit temperature indication and 16-bit humidity indication. A measurement can be requested every second with a non-interruptible 5ms blocking function. The sensor is calibrated and does not require any additional components to be used. Figure 4.13 shows the DHT22 humidity and temperature sensor and Figure 4.14 illustrates the positions adopted by the DHT22 sensors on the system under consideration (see Appendix 34).

Figure 4.13. DHT22 temperature and humidity sensor.

Figure 4.14. Temperature and humidity monitoring.

2.4 Acquisition systems

2.4.1 How it works

In the first experiment, all the sensors considered are connected to an Arduino Uno-type electronic control board. The data is then collected and stored by the Bluetooth module. Figure 4.15 (a) illustrates the wiring of the acquisition system we have developed. The DS18B20 temperature sensor is a 3-wire digital temperature sensor. Each sensor has a unique 64-bit serial number, which means that we can use several sensors on the same data bus. In other words, several sensors are connected to the same Pin on Arduino. This is particularly useful for data logging and temperature monitoring projects. The DS18B20 is an excellent sensor because it is inexpensive, accurate and very easy to use. Figure 4.15 (b) shows the connections between the components that make up the acquisition system. (See Appendix28)

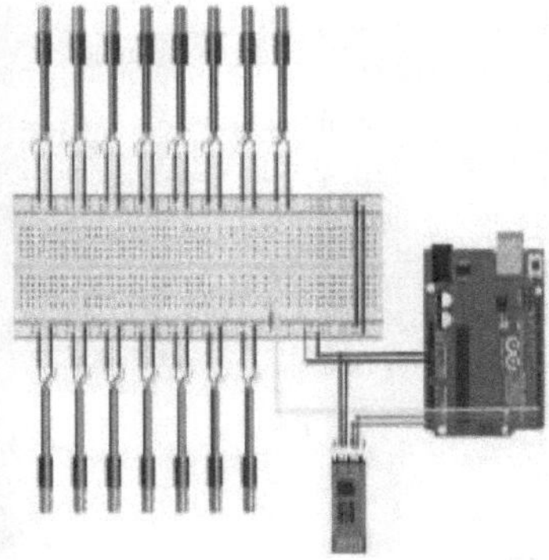

(a) Acquisition system (b) Schematic of the system via the Circuit interface.

Io

Figure 4.15. First experiment.

For the second experiment, the sensors are also connected via the Arduino Uno electronic control board. Figure 4.16 (a) illustrates the wiring of the acquisition system we have developed. The sensor in the DHT22 series can be used to measure the humidity and temperature of the ambient air using Arduino software. The DHT22 costs about twice as much as the DHT11, but will offer you a wider measurement range and greater accuracy. In other words, several sensors are connected to the same Pin on Arduino. This is particularly useful for data logging and temperature monitoring projects. Figure 4.16 (b) shows the connections between the elements making up the acquisition system. (See Appendix 29)

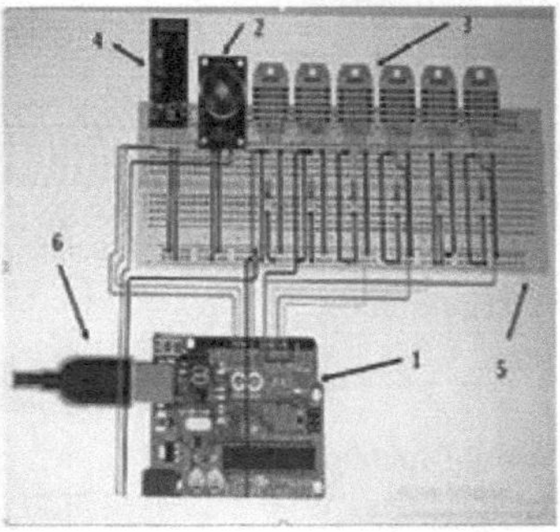

(a) Acquisition system (b) Electrical circuit diagram

N°	Description
1	Arduino Uno board
2	CO_2sensor
3	DHT22 temperature and humidity sensor
4	Bluetooth module HC-05
5	Test plate
6	USB cable

Figure 4.16. Second experiment

2.4.2 Data collection

2.4.2.1 HC05 Bluetooth module

The HC-05 Bluetooth module is used to establish a Bluetooth link (serial link) between an Arduino board and other equipment with a Bluetooth connection (smartphone, tablet, PC, second Arduino board, etc.). Figure (4.17) shows the HC-05 Bluetooth module.

This serial link is established on two pins RX and TX defined in our program as pins 11 and 10. The RX pin of the Uno-compatible board must be connected to the TX pin of the HC05 Bluetooth module and the TX pin of the Uno-compatible board must be connected to the RX pin of the HC-05 module.

Figure 4.17. HC-05 Bluetooth module.

Table (4.5) therefore describes all the characteristics of the Bluetooth module:

Table 4.5. Characteristics of the HC-05 Bluetooth module

Function	master / slave
Operating voltage	3.3V
Input voltage	3.3 to 6V.
Bluetooth standard protocol	V2.0
Default baud rate	9600, user modifiable
Size	39 x 15 mm
Operating current	30mA
Use for communication	via Bluetooth with mobile phones, tablets and computers.
Use of a regulator	150 mA 3.3V
6-pin pin	Vcc, Gnd, TxD, RxD

2.4.2.2 Serial Bluetooth Terminal Interface

The Serial Bluetooth Terminal application is a line-oriented terminal/console application for microcontrollers, Arduino and other devices with a serial/UART interface connected with a Bluetooth to serial converter to your Android device.

Figure 4.18 (a) shows the graphical interface of the "Serial Bluetooth Terminal" terminal emulator in the first experiment. Thanks to this application, Bluetooth connections can easily be established between the device and the PC to monitor the system's progress over time.

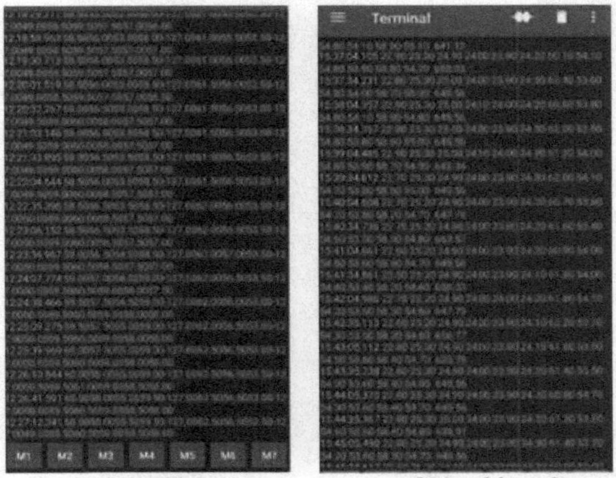

(a) Expérience 1. (b) Expérience 2.

Figure 4.18. Serial Bluetooth Terminal interface

3. Experimental study of thermal comfort

Thermal comfort in its different modes, natural ventilation and air conditioning, is probably the first thing everyone thinks of when they talk about comfort in buildings. The main physical parameters that govern the sensation of thermal comfort are temperature and humidity. In our experiment we are interested in the distribution of temperature and humidity in a room. In our case, the DHT22 will be used as an instrument to measure temperature and humidity .one of the sensor connected to the proximity of the air conditioner and the five DHTT22 connected to half of the room.

3.1 Monitoring over the course of a day

In this paragraph, we are interested in processing the results relating to four days, the first day being 26 June 2021 when this natural ventilation mode was used. The rest of the days, 28, 29 and 30 June 2021, correspond to the air-conditioning mode.

3.1.1 Natural ventilation mode

3.1.1.1 Instantaneous temperature in the DHT22 sensors

Figure 4.19 shows the instantaneous evolution of the temperature of the six DHT22 sensors in a room in natural ventilation mode during the first day of 26 June 2021. From these results, it can be seen that the temperature in the different positions of the DHT22 sensors increases, then decreases, with the highest value occurring between 12 noon and 2 pm. At 8am, the temperature was between T=28°C and 29°C. After this time, there was a gradual increase in temperature until a maximum value of T=31.5 °C was reached. After this increase in temperature, there was a continuous, non-acute decrease in temperature until the minimum value of T=28.5 °C was reached at 4 p.m. for positions 3 and 4. For the rest of the day, at 5 p.m., the temperature increased slightly and varied between 30 and 30.5 °C.

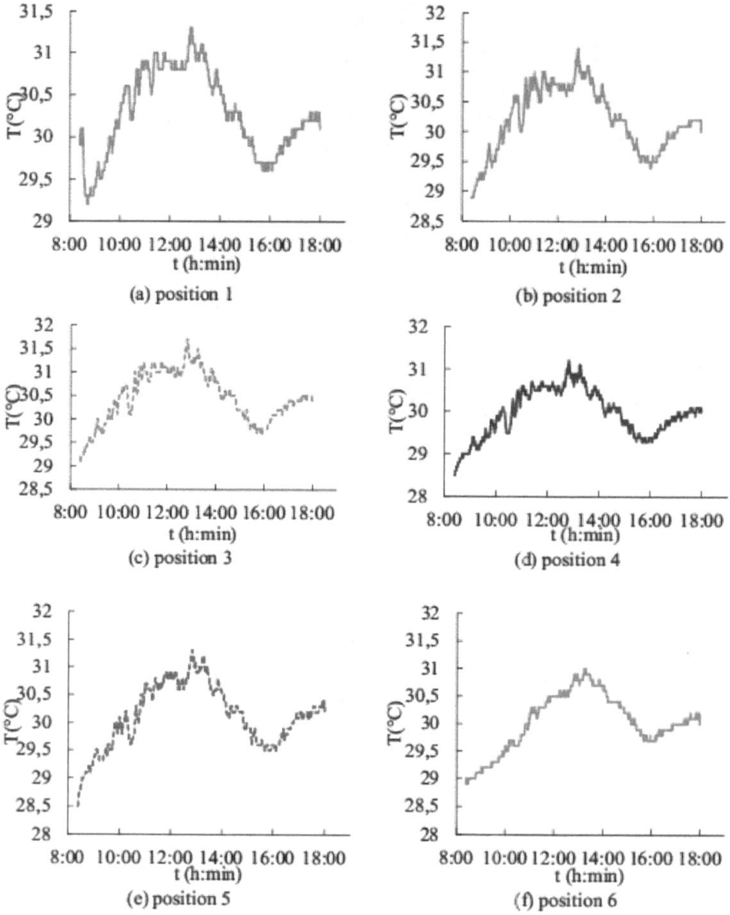

Figure 4.19. Hourly temperature variation for the six DHT22 sensors on 26 July 2021.

3.1.1.2 Average temperature

Figure 4.20 shows the instantaneous evolution of the average temperature of the six DHT22 sensors in the chamber during the day of 26 June 2021. From these results, we can see that the temperature in the different positions of the DHT22 sensors is increasing. It then decreases between 10 am and 2 pm. At 9am, the temperature was equal to T=29°C. After this time, there was a gradual increase in temperature until a maximum value of T=31.3 °C was reached. After this increase in temperature, there was a continuous, non-acute decrease in temperature until the value of T=29.5 °C was reached. For the rest of the day, the temperature is equal to 30°C between 5 p.m. and 6 p.m.

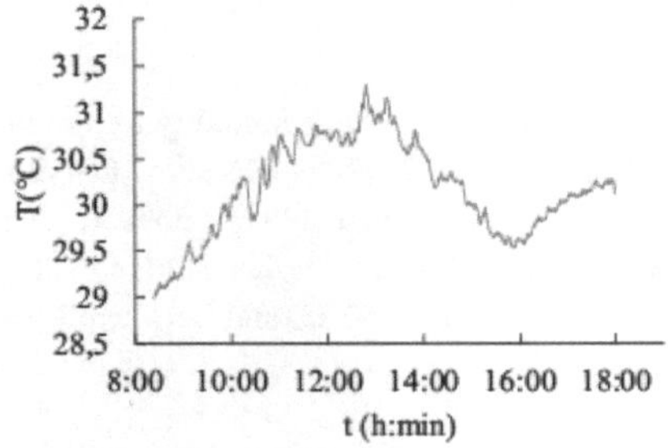

Figure 4.20. Variation in mean temperature for the six sensors on 26 June 2021.

3.1.1.3 Humidity in the DHT22 sensors

Figure 4.21 shows the hourly variation in relative humidity for the day of 26 June 2021 in natural ventilation mode. These results show that the relative humidity shows high values in the morning with a humidity value equal to H=72.9%. As the sun rises, relative humidity gradually decreases until 1pm. It then increased again, reaching higher values between 2pm and 3pm. At 4 p.m. it decreases until it reaches the value H=45% and at 6 p.m. it resumes this increase, which corresponds to sunset.

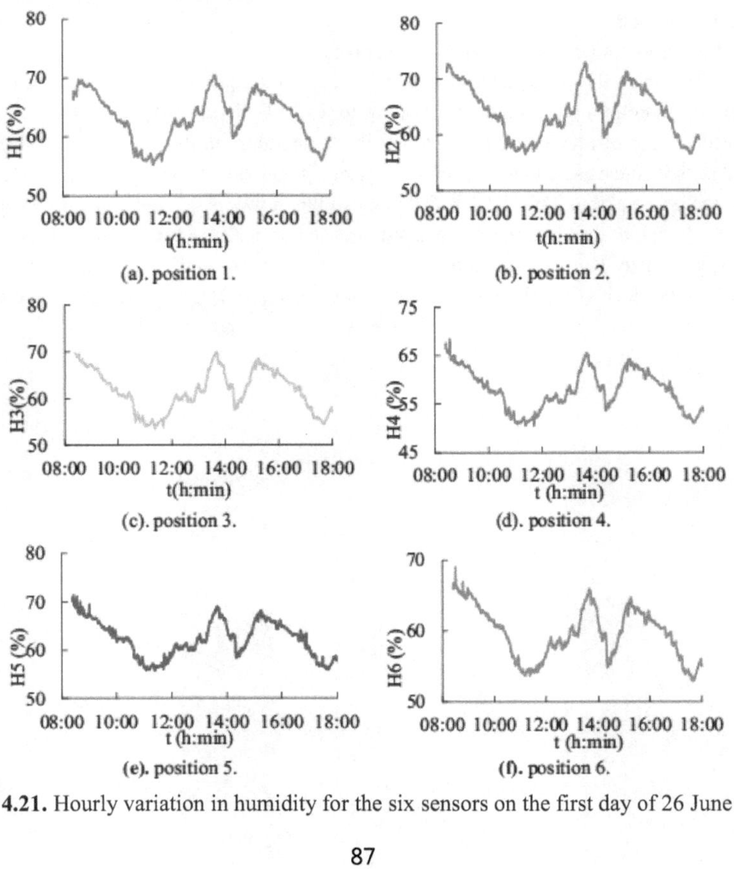

Figure 4.21. Hourly variation in humidity for the six sensors on the first day of 26 June

2021.

3.1.1.4 Average humidity

Figure 4.22 shows the hourly variation in average humidity for the first day of the experiment in June 2021 in natural ventilation mode. These results show that the relative humidity is low in the morning, with a humidity value equal to H = 55% 1. The relative humidity then increases progressively until 14:00 at H = 69%. It then starts to fall again until it reaches low values between 2 and 3 p.m. equal to H = 57°C, and between 3 and 6 p.m. it starts to rise again.

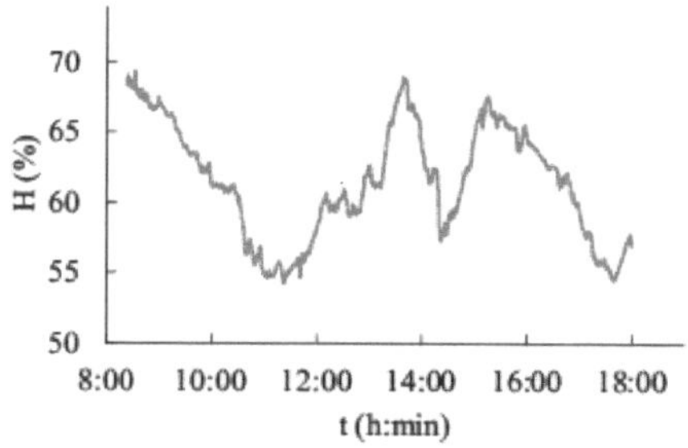

Figure 4.22. Variation in average humidity for the six sensors on 26 June 2021.

3.1.2 Cooling mode

3.1.2.1 Instantaneous temperature in the sensors

Figure 4.23 shows the instantaneous change in temperature of the six DHT22 sensors in a room in air-conditioning mode corresponding to T=20°C during the second day of 28 June 2021. From these results, it can be seen that the temperature in the different positions of the DHT22 sensors decreases, then becomes constant. The highest temperature value occurred at 8 a.m., and was equal to T=29.9°C in DHT sensor position 3. After this time, the temperature gradually decreased until it reached a maximum value of T=31.5°C. After this increase in temperature, there was a continuous, non-acute decrease until the minimum value of T=20.7°C was reached for the first sensor in positions 3 and 4. For the rest of the day, the temperature remained constant, varying between T=21.5°C and T=25.5°C.

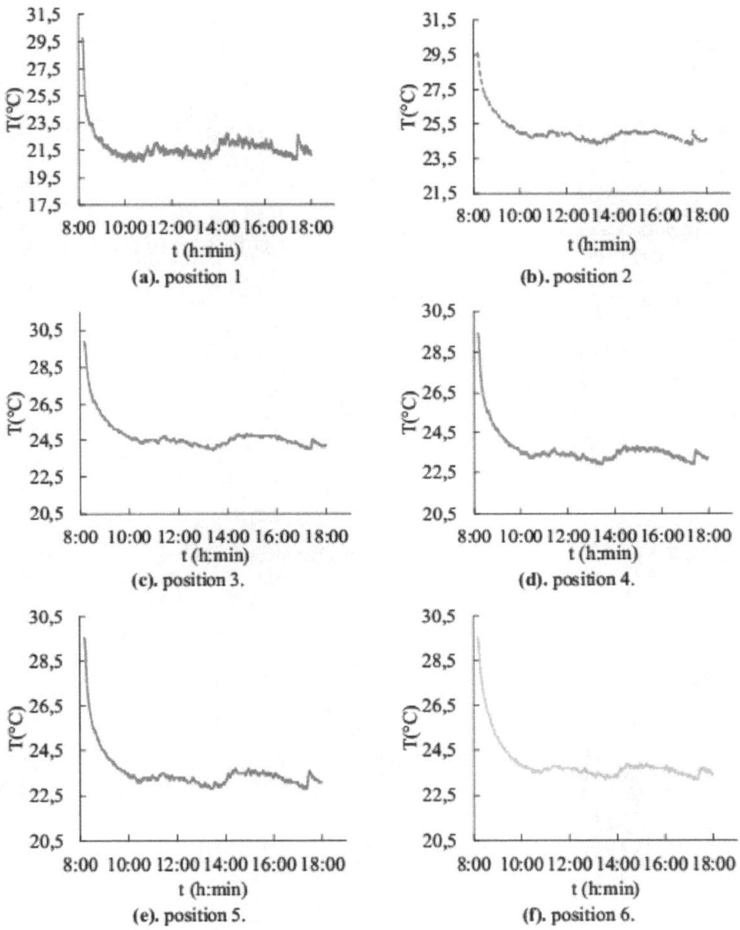

Figure 4.23. Hourly temperature variation for the six DHT22 sensors on 28 July 2021.

Figure 4.24 shows the instantaneous temperature variation of the six DHT22 sensors in a room in air-conditioning mode corresponding to T=26°C during the third day, 29 June 2021. From these results, it can be seen that the temperature in the various positions of the DHT22 sensors decreases from 8 o'clock in the morning. After this time, there were two peaks in temperature at 9am and 12pm, reaching a maximum value of T=27.7°C in the sensor at the first position. After this increase in temperature, there was a continuous non-acute decrease in temperature until the minimum value of T=22.5°C was reached for the first sensor in the first position and at 5 p.m. Apart from the presence of the peak, there was a decrease for the rest of the day.

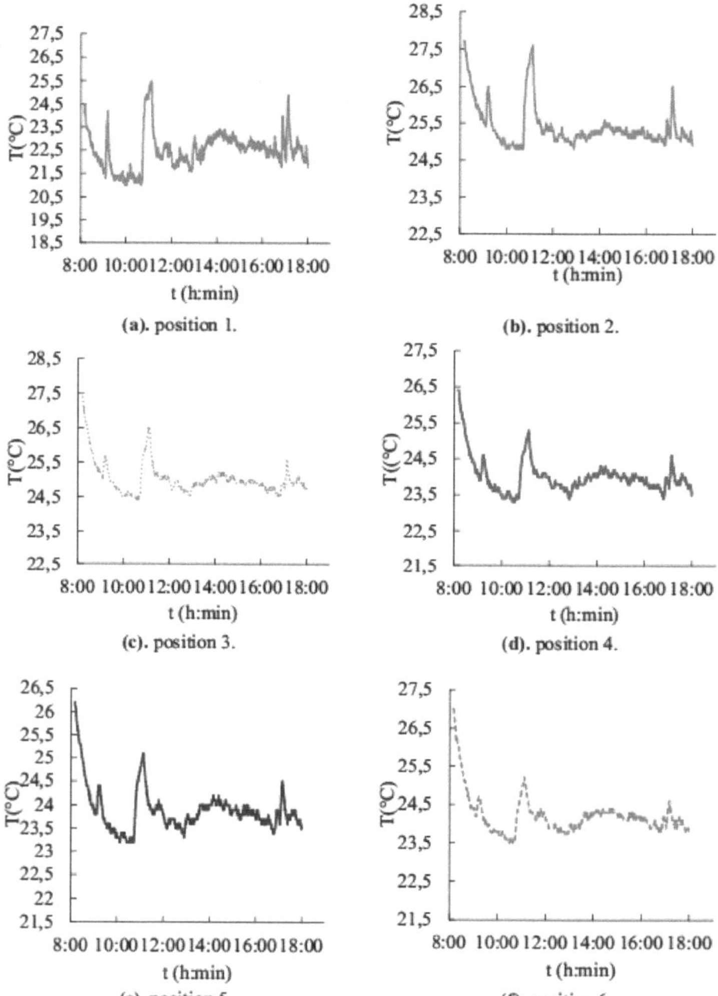

Figure 4.24. Hourly temperature variation for the six DHT22 sensors for the day of 29 July 2021.

Figure 4.25 shows the instantaneous temperature variation of the six DHT22 sensors in a room in air-conditioning mode at T=32 °C during the fourth day of 30 June 2021. From these results, it can be seen that the temperature in the different positions of the DHT22 sensors increases, then remains constant, with the highest value occurring at 8 o'clock in the morning and varying between T=25.5 °C and 26.5 °C. After this time, there is an increase in the temperature of the DHT22 sensors. After this time, the temperature gradually increased until it reached a maximum value of T=30.5°C for the rest of the day. Furthermore, there was no difference in temperature between the six DHT22 sensors.

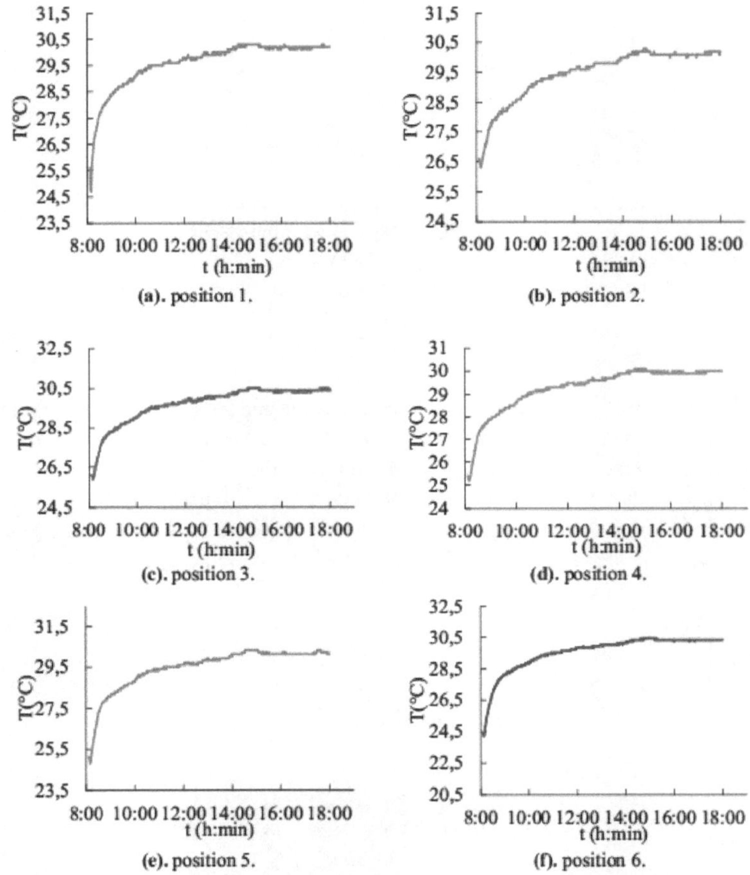

Figure 4.25. Hourly temperature variation of the six DHT22 sensors for the day 30 July 2021.

3.1.2.2 Average temperature

Figure 4.26 shows the instantaneous change in the average temperature of the six DHT22 sensors in a room during the day of 28 June 2021 at T = 20°C. From these results, it can be seen that the temperature in the different positions of the DHT22 sensors decreases and then remains constant. At 8am, the highest temperature value was T=29.5°C. After this time, the temperature gradually decreases until it reaches a minimum value of T=21.5°C at 10 am. For the rest of the day, the temperature varied between T=22.5°C and T=24.5°C between 5 p.m. and 6 p.m.

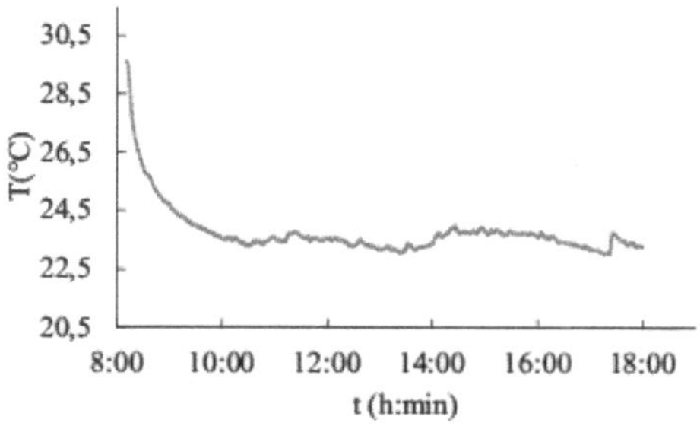

Figure 4.26. Variation in mean temperature for the six sensors on the second day of 28 June 2021.

Figure 4.27 shows the instantaneous variation in the average temperature of the six DHT22 sensors in the chamber during the day of 29 June 2021 at T=26°C. From these results, it can be seen that the temperature in the different positions of the DHT22 sensors varied between T=24.5°C and T=26.5°C. The temperature peaks at 12 o'clock until it reaches a maximum value of T=25.5°C. After this increase in temperature, there is a continuous, non-acute decrease in temperature until it reaches T=24.5°C for the rest of the day.

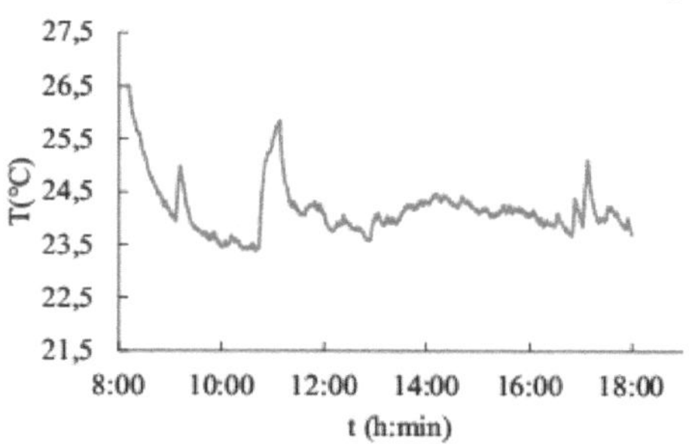

Figure 4.27. Variation in the mean temperature of the six sensors for the day of 29 June 2021.

Figure 4.28 shows the instantaneous variation in the average temperature of the six DHT22 sensors in the chamber during the day of 30 June 2021. From these results, it can be seen that the temperature in the different positions of the DHT22 sensors is constant. At 8 o'clock in the morning, the temperature is equal to T=25.5 °C. It then increases progressively until it reaches a value between T=29.5 °C and T=30.5 °C for the rest of the day.

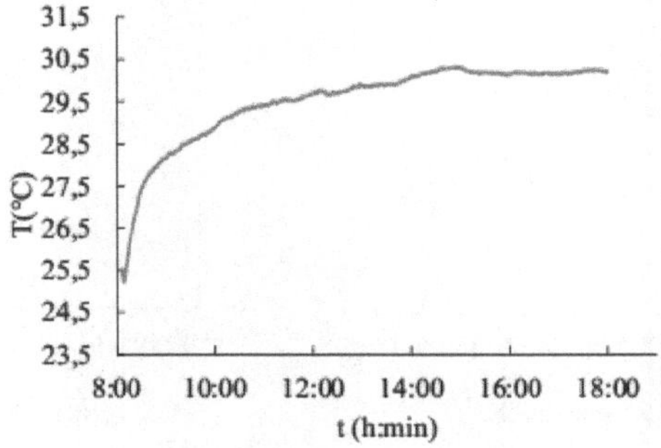

Figure 4.28. Variation in the average temperature of the six sensors for the day of 30 June 2021.

3.1.2.3 Humidity in the DHT22 sensors

Figure 4.29 shows the hourly variation in relative humidity for the second day of 28 June 2021 in air-conditioning mode T=20°C. These results show that the relative humidity shows high values in the morning with a humidity value equal to H= 73.3% for the position of sensor1. From this time onwards, relative humidity fell gradually until 2pm. It then increased again, reaching high levels between 2pm and 3pm, equal to H= 67°C. At 4pm, it fell until it reached the minimum value of H= 46% for the sensor at position 2. At 18 h, it resumes this increase.

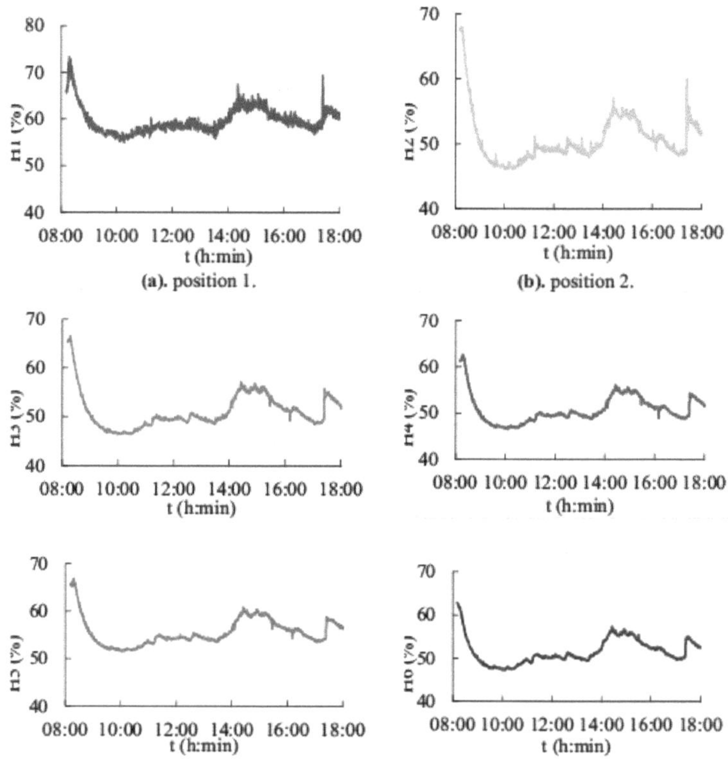

Figure 4.20. Hourly variation in humidity for the six sensors on 28 June 2021.

Figure 4.30 shows the hourly variation in relative humidity for the day of 29 June 2021 corresponding to air-conditioning mode T= 26 °C. These results show that the relative humidity shows high values in the morning with a humidity value equal to H= 72.2 % for the sensor at position 1. It then increased again, reaching high values between 2pm and 4pm, equal to H=65°C. Then, at 4pm, it fell until it reached a minimum value of H=52% for the sensors in positions 2 and 6. At 5 p.m. the humidity value starts to rise again.

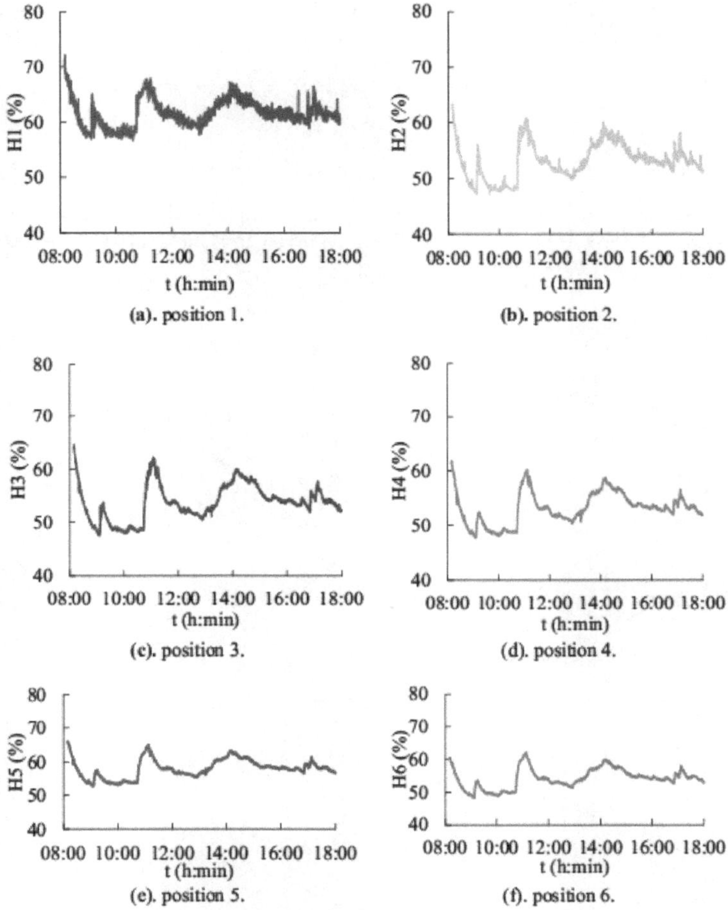

Figure 4.30. Hourly variation in humidity for the six sensors on 29 June 2021.

Figure 4.31 shows the hourly change in relative humidity for the day of 30 June 2021 in air-conditioning mode, corresponding to T= 32 °C. These results show that relative humidity is low in the morning, with a humidity value equal to H= 49% for the sensor at position 1.

The relative humidity then increases gradually until it reaches H= 85% at 2pm for the sensor in position 2. It then began to fall again, reaching low values between 2pm and 3pm, equal to H=70% for the sensor at position 6. At 6 p.m. the humidity value starts to rise again.

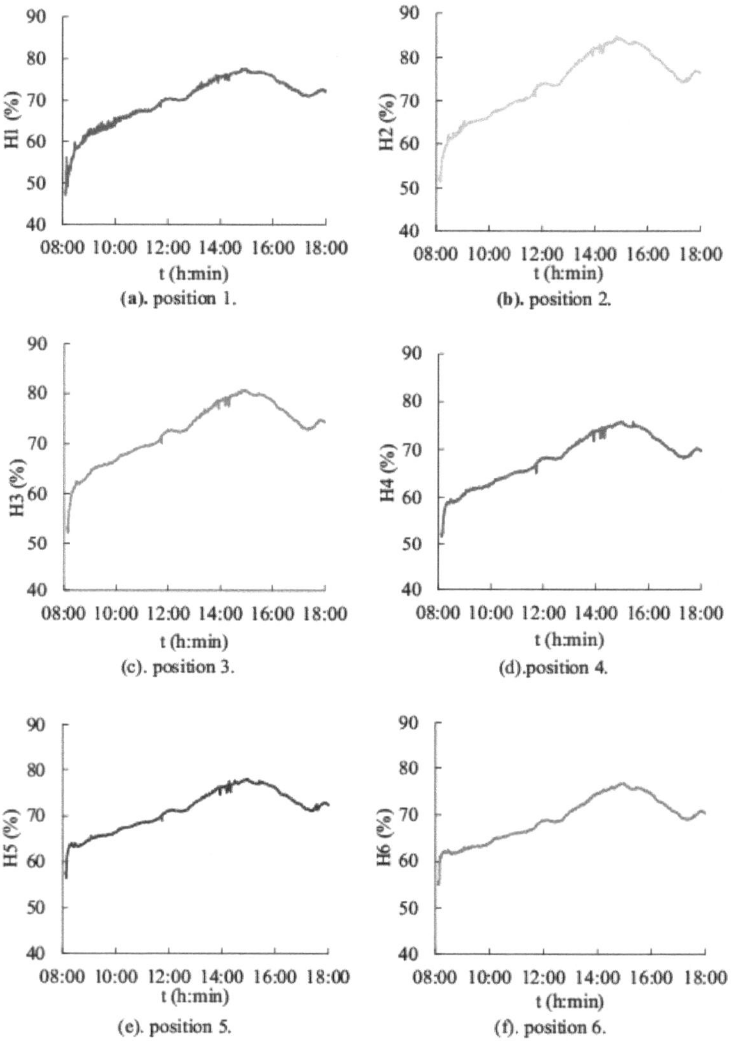

Figure 4.31. Hourly variation in humidity for the six sensors on the fourth day 30 June 2021.

3.1.2.4 Average humidity

Figure 4.32 shows the hourly variation in average humidity for the second day of the experiment in June 2021 in air-conditioning mode T= 20°C. These results show that relative humidity shows high values in the morning with a humidity value equal to H= 65%. Then the relative humidity gradually decreases until 2pm at H= 50%. It then increases again until it reaches values between 2 p.m. and 5 p.m. equal to H=53°C, and at 5 p.m. it increases again.

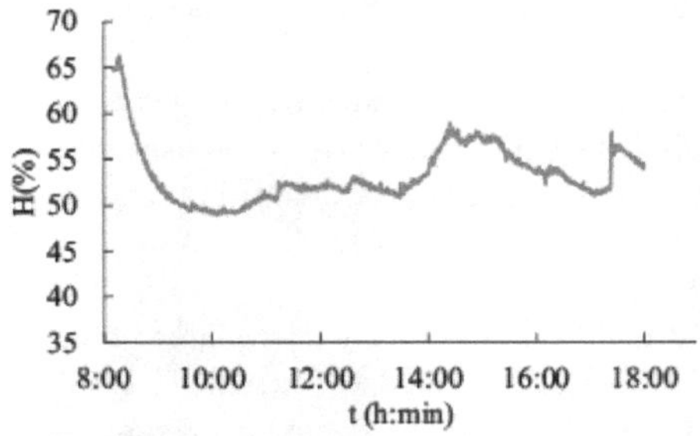

Figure 4.32. Variation in average humidity for the six sensors on 28 June 2021.

Figure 4.33 shows the hourly variation of the average humidity for the four days of the experiment in June 2021 in air-conditioning mode T=26°C. These results show that the relative humidity has high values in the morning with a humidity value equal to H=65%. The relative humidity then fell gradually until 11am, at H= 53% a. It then peaks and increases again, reaching high values between 12 noon and 2 p.m., equal to H=63%, and for the rest of the day, at 5 and 6 p.m., it continues to rise.

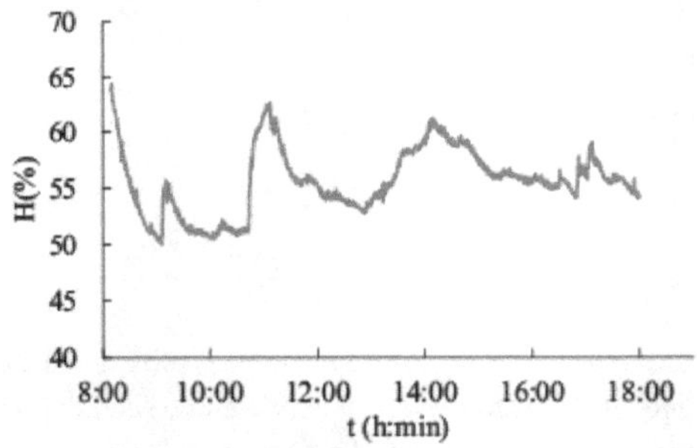

Figure 4.33. Variation in average humidity for the six sensors on 29 June 2021.

Figure 4.34 shows the hourly variation in mean humidity for the fourth day of the experiment in June 2021 in air-conditioning mode T=32°C. These results show that relative humidity is low in the morning, with a humidity value equal to H=52%. The relative humidity then rose gradually until 14:00, when it reached H=79%, then fell again until it reached low values between 14:00 and 15:00, equal to H=70°C, and then rose again at 18:00.

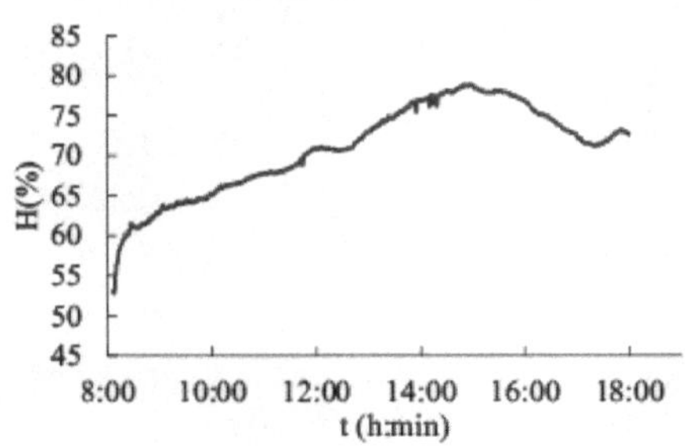

Figure 4.34. Variation in mean humidity for the six sensors on 30 June 2021.

3.2 Temperature and mean humidity as a function of the z point

In our experiment, we fixed six DHT22 sensors linked to the acquisition system in one half of the room in order to monitor its temperature and humidity along the direction (oz) with the x and y axes fixed for four days, 26, 28, 29 and 30 June 2021.

3.2.1 Natural ventilation mode

Figure 4.35 shows the evolution of the average temperature of the different sensor positions during the day of 26 June 2021. From these results, it can be seen that the daytime temperatures of the six DHT22 sensor positions are almost identical apart from a few disturbances. At z=1.5 m, an increase in the average temperature is observed, reaching a maximum value of 30.46°C. The same figure also shows the characteristic equation T=f (z), which is determined from two variable parameters: the mean temperature T_{avg} and the z point.

$$Tmoy = -0{,}1114x + 30{,}375 \quad (4.1)$$

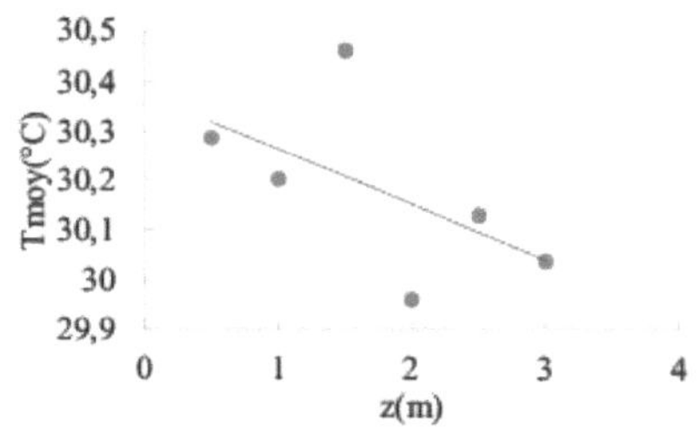

Figure 4.35. Mean temperature as a function of z for 26 June 2021.

Figure 4.36 shows the evolution of the average humidity of the different sensor positions during the day of 26 June 2021. From these results, it can be seen that the humidity of the day at the six DHT22 sensor positions is almost identical apart from a few disturbances. At z = 1 m the average humidity increases until it reaches a maximum value of H_{avg}= 64 %, then at z = 1.5 m and 2 m it decreases to an equal value of H_{avg}= 58 %, and between 2 m and 3 m it increases again. The same figure also shows the characteristic equation H=f (z) which is determined from two variable parameters: the mean temperature H_{avg} and the z point.

$$H = -1{,}5211x + 64{,}107 \quad (4.2)$$

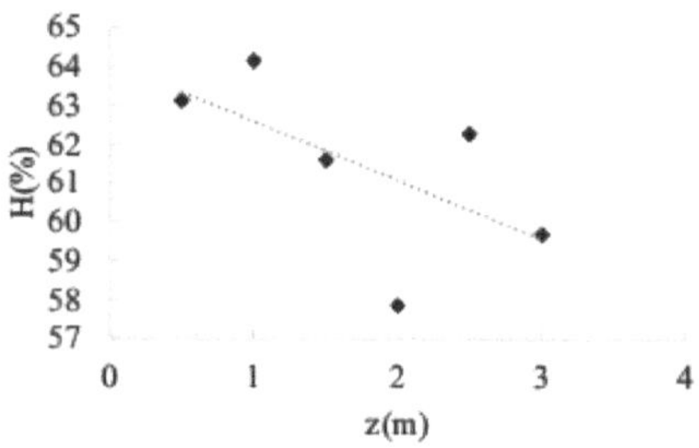

Figure 4.36. Average humidity as a function of z for 26 June 2021.

3.2.2 Cooling modes at T=20°C

Figure 4.37 shows the curve of the mean temperature as a function of the height z in m for the day 28 June 2021.Thus, we have determined the characteristic equation for Tmoy, which can be found in the following equation :

$$T\ moy = -0{,}6774x + 25{,}525 \quad (4.3)$$

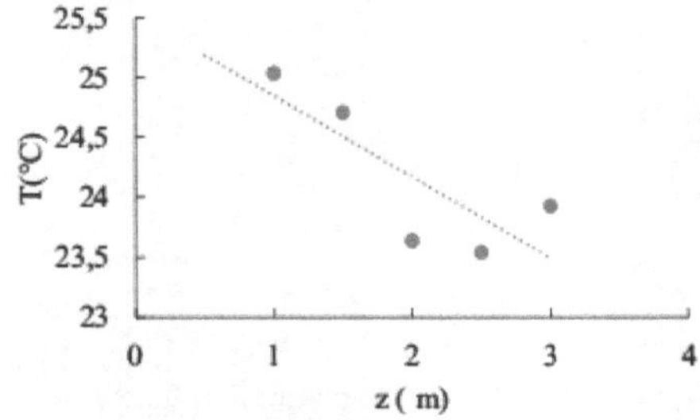

Figure 4.37. Mean temperature as a function of z for 28 June 2021.

Figure 4.38. Shows the average humidity in a closed room for the day of 28 June 2021, thus the characteristic equation T=f(z).

$$H\ moy = 1{,}2525x + 49{,}469 \quad (4.5)$$

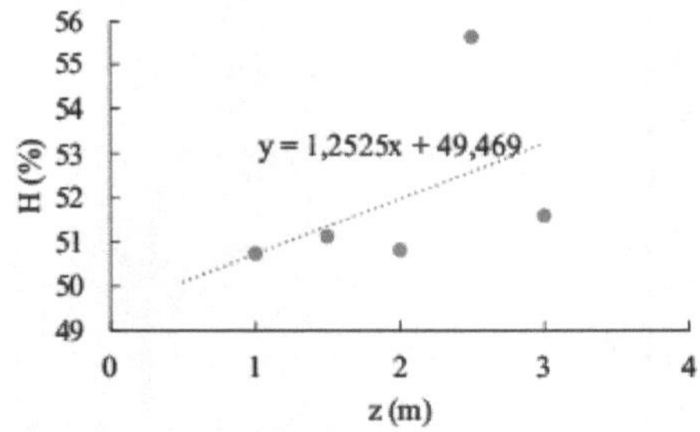

Figure 4.38. Average humidity as a function of z for 28 June 2021.

3.2.3 Cooling modes at T=26°C

Figure 4.39 shows the characteristic equation for the mean temperature Tmoy as a function of z for the day of 29 June 2021.

T moy= -0,7067x + 25,922 (4.6)

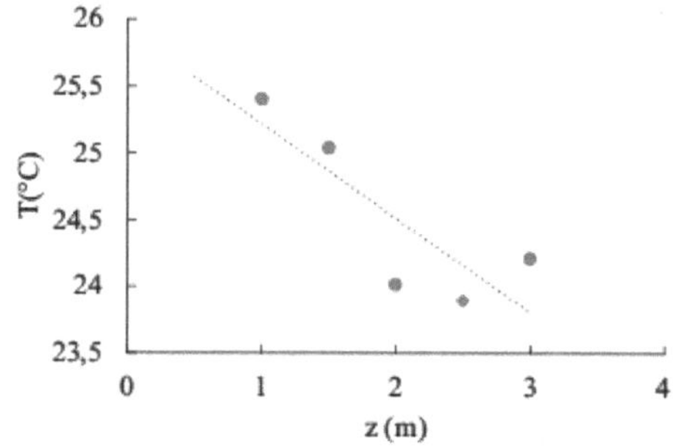

Figure 4.39. Mean temperature as a function of z.

Figure 4.40 shows the characteristic equation for mean humidity H mean as a function of z for the day of 29 June 2021.

H moy= 1,277x + 51,909 (4.7)

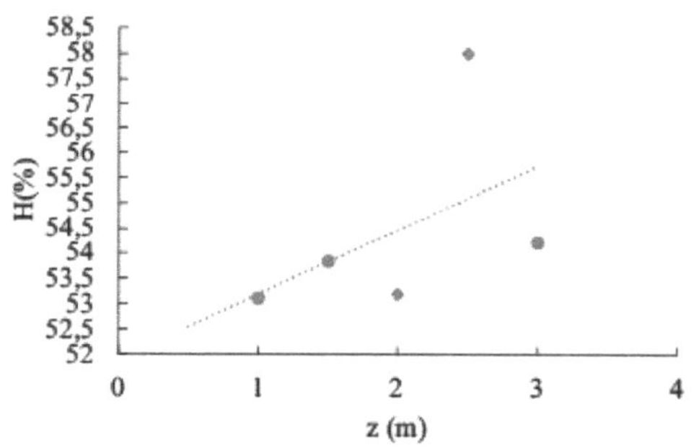

Figure 4.40. Average humidity as a function of z

3.2.4 Cooling modes at T=32°C

Figure 4.41 shows the characteristic equation for the mean temperature Tmean as a function of z for the day of 30 June 2021.

T moy= -0,0409x + 29,552 (4.8)

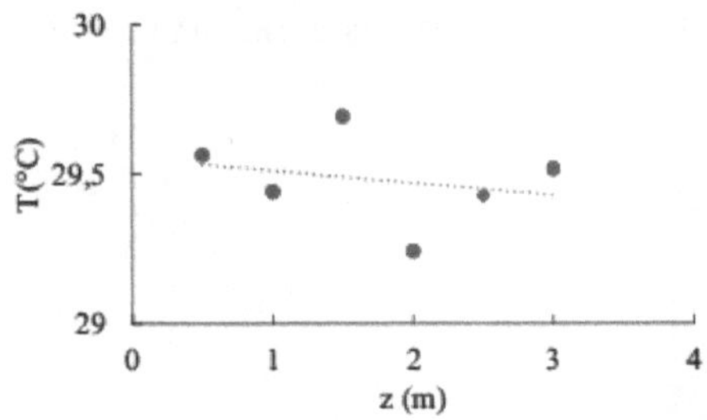

Figure 4.41. Average temperature as a function of z for the day of 30 June 2021.

Figure 4.42 shows the characteristic equation for mean humidity Hmoy as a function of z for the day of 30 June 2021.

y = -0,6545x + 71,299 (4.9)

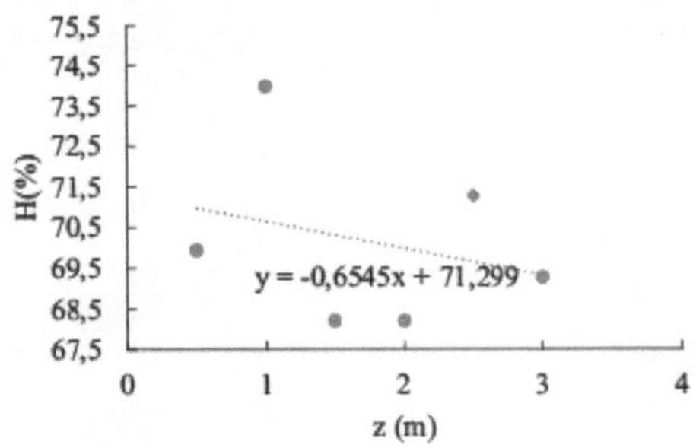

Figure 4.42. Average humidity as a function of z

3.2.5 Correlation between temperatures in air-conditioning mode

In our experiment in the laboratory over 4 days at the end of June 2021, the idea is to study the correlation between the temperature and the humidity.

for our experiment is to study the correlation between the temperatures in air-conditioning mode at

T=20°C, 26°C and 32°C in order to determine the new generalized equation of the temperature at the point z when the air conditioner is set to a certain mean temperature and at some value of z in m.

With the co-ordinates of the point at the median line in the room with respect to x and y but to the variation in z.

So two curves are drawn, the first curve is the average temperature (20°C, 26°C and 32°C) as a function of the first coefficient T0 as found in the following equations.

For T=20 ° we have

T = f(z) y = -0.6774x + 25.525

For T=26 °C

T = f(z) y = -0.7067x + 25.922

For T=32 °C

$$T = f(z)\ y = -0.0409x + 29.552$$

Figure 4.43 shows the mean temperature curve (20; 26; 32°C) as a function of the T0 terms (-0.6774; -0.7067; -0.0409). The characteristic equation is thus defined as follows:

$$T\ moy = 0{,}0224x - 1{,}\ 1792 \quad (4.10)$$

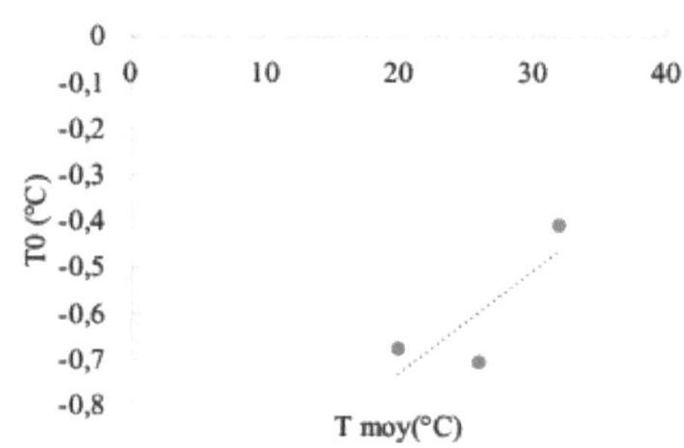

Figure 4.43. Average temperature as a function of the coefficient T0

Figure 4.44 shows the average temperature curve (20°C; 26°C; 32°C) as a function of the T1 terms (25.525°C; 25.992°C; 29.552°C). The characteristic equation is thus defined as follows:

$$y = 0{,}3356x + 18{,}274 \quad (4.11)$$

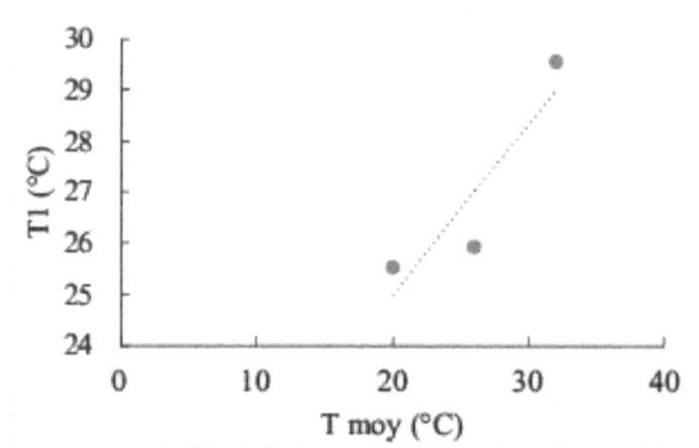

Figure 4.44. Mean temperature as a function of the second coefficient T1

Therefore, the new general equation which determines the point temperature as a function of z for the several two-parameter conditions T and z can be found in equation (4.12)

$$T_z = (0{,}0224\ Tmoy - 1{,}\ 1792)\ z + (0.3356T_{moy} + 18{,}274) \quad (4.12)$$

4. Experimental study of the photovoltaic system

In this work, we develop an experimental study focusing on the characterization of a photovoltaic solar panel that is available in the Laboratoire des Systèmes Electromécaniques (LASEM) of the Ecole des Ingénieurs de Sfax (ENIS).

In this experiment we are interested in following the electrical production of the solar collector fixed at an inclination of 30° and a southern orientation at the beginning of July, the first day is to follow the production of the collector without cooling and the second and third days the collector with cooling every two hours and every half hour respectively by water runoff for each day of 0.5 ml.

(a) panel without cooling(b) panel with cooling

Figure 4.45. PV sensor operating mode

4.1 Environmental parameters

For the study of the effect of the variation of the meteorological data on the operation of photovoltaic solar collector studied, we referred to different days of experiments; namely 01 July 2021, 02 July 2021 and 03 July 2021.

4.1.1 Sunshine

Figure 4.46 shows the hourly sunshine values at the sensor plane in the Sfax region corresponding to a clear sky, with the measurement time interval for each day varying from 8 a.m. to around 6 p.m. The distribution of sunshine at the surface of the photovoltaic module for three days in July 2021 is almost equal. The distribution of sunshine on the surface of the photovoltaic module for three days in July 2021 is almost equal. These results show that solar radiation is low in the morning. These values increase as the sun rises, reaching their highest values between 10 and 11 a.m. Thereafter, the drop in sunshine for the first and second day at 12 noon is due, exceptionally, to the passage of clouds. From 1pm onwards, there was an increase in sunshine, reaching 980 $W.m^2$for the first day, and from 2pm onwards the solar radiation became less intense, reaching low values at the end of the day of between 900 $W.m^2$and 120 $W.m^2$. Under these conditions, the highest value of solar intensity is equal to E=1008 $W.m^{-2}$which is obtained for the second day of 02
July 2021 at 11:20 h.

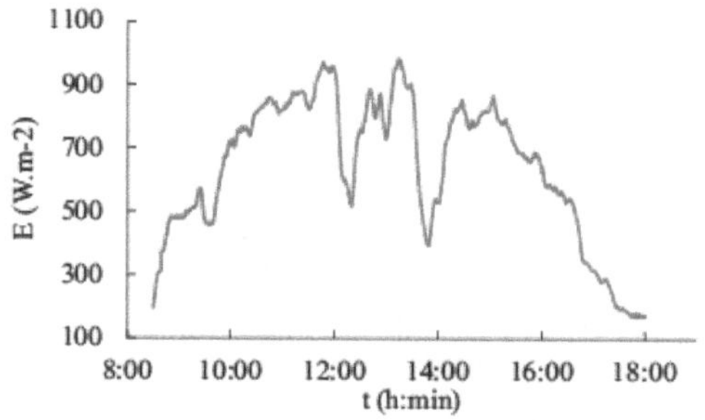

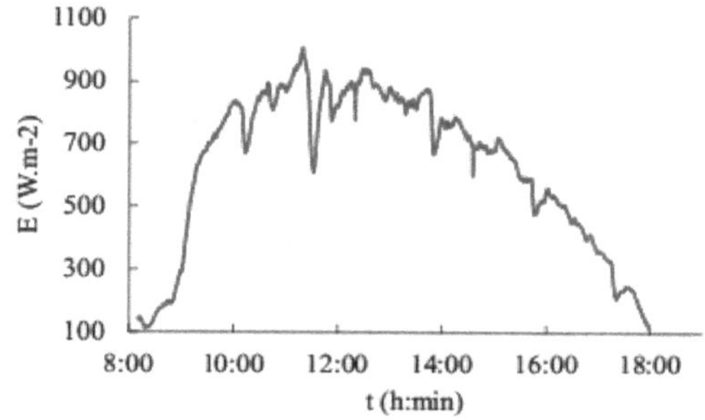

(b). capteur PV avec refroidissement chaque deux heures

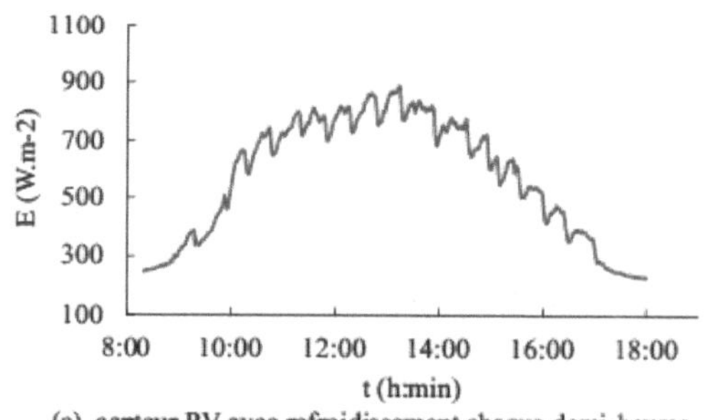

(c). capteur PV avec refroidissement chaque demi-heures

(b). PV collector with cooling every two hours

(c). PV collector with cooling every half hour

Figure 4.46. Sunshine received for three days from the photovoltaic collector

4.1.2 Ambient temperature

Figure 4.47 shows the profile of the instantaneous variation in ambient temperature measured by thermocouple No. 9 on 01, 02 and 03 July 2021. On 01 July the ambient temperature varied between 34°C and 42°C, on 02 July the variation was between 31°C and 44°C and on 03 July 2021 the ambient temperature varied between 30°C and 39°C. The temperature distribution on the three days was almost the same. Under these conditions, on the first day, the rise in temperature was recorded between 12pm and 3pm. On the second day, the temperature increase was recorded between 11am and 3pm, and on the last day, the temperature increase was recorded between 10am and 2pm. As a result, the large instantaneous fluctuation in the temperature profile is due to the extreme change in the intensity of solar radiation.

The first day is carried out for a PV collector in uncooled mode, and the results show that the ambient temperature is low in the morning. As the sun rises, the ambient temperature gradually increases until 11 a.m., then decreases until 1 p.m. due to the presence of clouds. From 1:30 p.m. until 3 p.m., the ambient temperature increases, then decreases again until reaching the lowest values at 6 p.m., which corresponds to sunset.

On the second day, for a PV collector operating in cooling mode every two hours, the ambient temperature was low from 7 am to 10 am. Then, from 11 am onwards, it began to rise to between 39°C and 44°C until 3 pm. From 4 p.m. it continues to fall until it reaches its lowest value at 6 p.m., equal to 31°C, which corresponds to sunset.

For the third day, for example, the PV sensor is operated in cooling mode every half hour, and at 7am the ambient temperature is low. It then began to rise on average between 10 and 15 o'clock until it reached a value of 39°C. From 4 p.m. it continues to fall until it reaches its lowest value at 6 p.m., equal to 30°C, which corresponds to sunset.

Under these conditions, the highest value of the ambient temperature is T_A= 44°C for the day of 02 July 221.

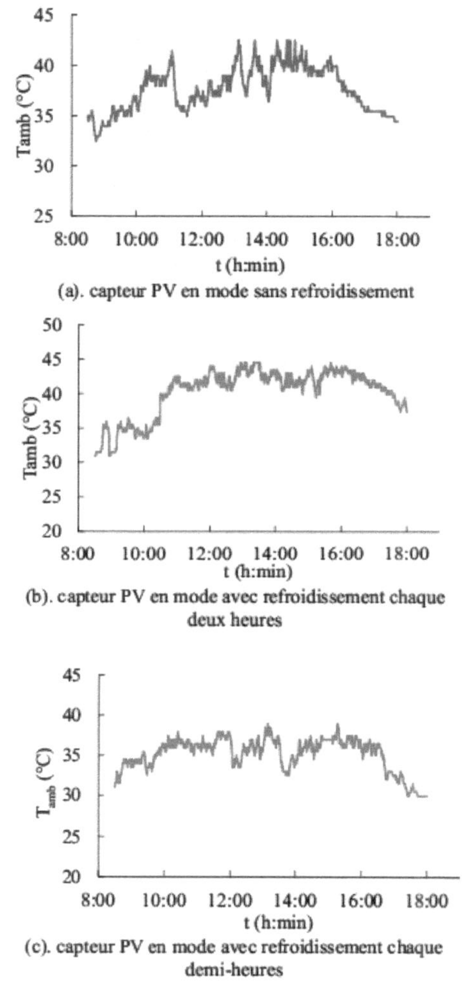

(a). PV collector in uncooled mode

(b). PV collector in cooling mode every two hours

(c). PV sensor in mode with cooling every half hour

Figure 4.47. Hourly variation in ambient temperature for three days in July 2021.

4.2 Daily monitoring

Daily monitoring consists of switching on the PV sensor for a whole day, collecting the data with a frame grabber, storing it and sending it via the Bluetooth module.

For the experimental study, we fixed our PV panel on a single 180° orientation (due south) and a single 30° inclination of this panel to monitor the variation in the panel's characteristics. The fifteen thermocouples installed on the photovoltaic collector enable the temperature of each solar collector cell to be monitored, except for sensor No. 9, which tracks the ambient temperature during the day.

4.2.1 Average temperature monitoring

The average temperature is calculated from fourteen thermocouples (n°1, n°2, n°3, n°4, n°5, n°6, n°7, n°8, n°10, n°11, n°12, n°13, n°14, n°15).

Figure 4.48 shows the instantaneous change in the average temperature of the solar photovoltaic collector on 01 July 2021, 02 July 2021 and 03 July 2021. From these results, it can be seen that the temperatures at the rear of the panel in the different positions of the DS18b20 collectors are almost comparable. However, under these conditions, the highest values correspond to the day of 03 July 2021, which took place between 12 noon and 3 pm. On this same day, at 7am, the sensor temperature was equal to T=31°C. After this time, there was a gradual increase in temperature until a maximum value of T=65°C was reached. After this increase in temperature and for the rest of the day, there was a continuous, non-acute decrease in temperature until it reached T=30°C at 6pm. This is explained by the decrease in solar radiation.

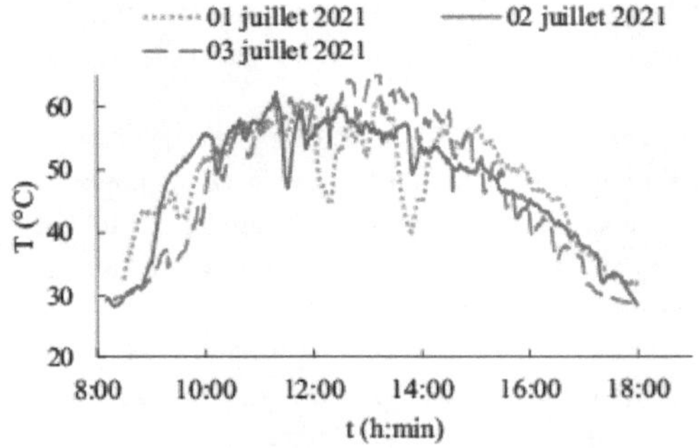

Figure 4.48. Instantaneous variation in mean temperature as a function of time.

4.2.2 Power monitoring

Figure 4.49 shows the instantaneous variation in load power on 01 July 2021, 02 July 2021 and 03 July 2021. From these results, it can be seen that the powers for the different days are almost comparable. The lowest values correspond to the day of the solar collector without cooling on 01 July 2021 between 12 noon and 2 p.m. and at 7 a.m., the powers are equal to P=47W. After this time, there is a gradual increase in power until a maximum value equal to P=210W is reached. After this increase, and for the rest of the day, we see a decrease until we reach a value of P=39W at 6pm. This is explained by the reduction in solar radiation. The two falls in power observed on 01 July 2021 at 12 noon and 2pm were due to the passage of a cloud.

As a result. It can also be seen that all the power values obtained for the last two days of the sensor with cooling every two hours and every half hour respectively are always greater and better than those obtained without cooling.

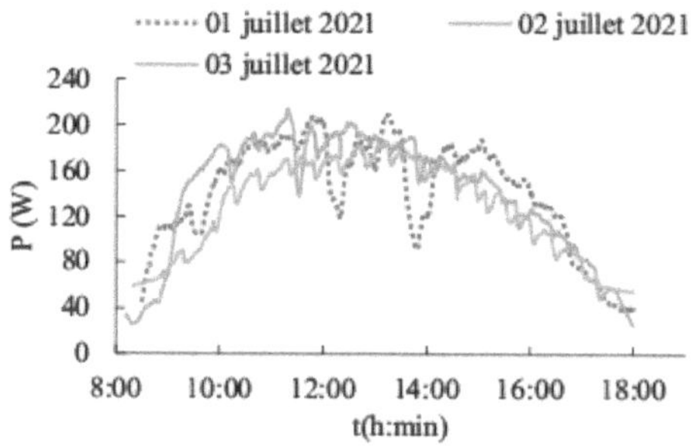

Figure 4.49. Instantaneous variation in power as a function of time.

4.2.3 Efficiency

The yield of photovoltaic solar panels is defined as the proportion of solar radiation that is transformed into electricity. The performance of a photovoltaic installation is characterised by its yield, taking into account the cooling of the solar panels, which has an influence on their productivity.

Figure 4.50 shows the evolution of yield for the three days 01, 02, and 03 July 2021. The first day is set for the panel without cooling and the last two days correspond to the PV collector with cooling every two hours and every half hour. From these results it can be seen that these PV collector efficiencies are not identical. It can be seen that for the first day at 7am, the efficiency is equal to η=14% and decreases until it reaches the value 13.5% at 11am. Then it gradually increases (two peaks) until it reaches a value equal to η=14.5% at 12 noon and 4 pm. At around 6pm, the yield starts to fall again.

Then, for the second day and at 9 a.m., the yield with the highest value is equal to η=14.35%, then it begins to decrease between 10 a.m. and 4 p.m. until it reaches a value equal to η=13.10%. From 5 p.m. onwards, the yield increases to a value equal to η=14.42%. Around 6 p.m., it starts to decrease again.

For the last day, at 7 a.m., the yield shows a high value equal to η=14.65%. It then decreases between 10 a.m. and 4 p.m. until it reaches a low value equal to 12.87%. From 5 p.m. onwards, there is an increase to a value equal to η=14.69%.

It can therefore be seen that the improvement in the efficiency of the cooled PV panel is evaluated by the relative difference in the efficiency of this panel compared with the situation of an uncooled panel.it is possible to improve the performance of the photovoltaic collector. In our case, the improvement is 0.63%.

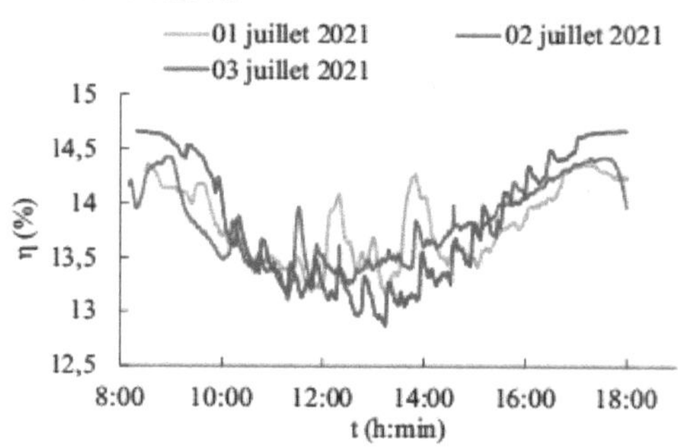

Figure 4.50. Variation in efficiency as a function of time.

4.3 Effect of thermocouple position

In our experiment, we implanted 14 thermocouples on the rear face of the PV sensor fixed at an inclination ß=30° at an orientation θ=180° in order to follow its temperature along three directions: 2 directions along the (ox) axis and one direction along the (ox) axis during three days 01,02 and 03 July 2021.

4.3.1 Temperatures along the (ox) axis

Figure 4.51 shows the evolution of the PV collector temperature for the day of 01 July 2021 in uncooled mode. From these results, it can be seen that the temperatures of the six positions of the DS18b20 sensors increase to reach an equal maximum value T=58.5°C at 11 o'clock. This temperature then decreased between 11am and 3pm. At 1pm, the temperature fell until it reached a minimum value of T=44°C at t=13h. At the end of the day, there was a continuous, non-acute decrease in temperature until reaching a value of T=37°C at 5 p.m.

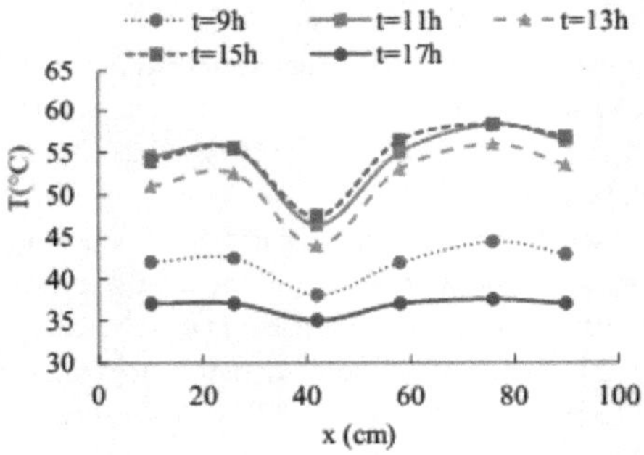

Figure 4.51. Variation in DS18b20 temperatures along the (ox) axis for the day on 01 July 2021

Figure 4.52 shows the evolution of the temperature of the PV collector for the day of 02 July 2021 carried out in PV collector mode with cooling every two hours. From these results, it can be seen that the temperatures of the various thermocouples present low intensities in the morning until reaching the minimum value T= 34 °C at 9 am. The temperatures of the six positions of the DS18b20 sensors then increased to reach an equal maximum value of T=62.5°C at 11am. This temperature then fell between 11am and 3pm. At 3pm, the temperature falls until it reaches a minimum value equal to T= 43.5°C at t=15h. At the end of the day, there was a continuous, non-acute decrease in temperature until the value T=36.5°C was reached at 5pm.

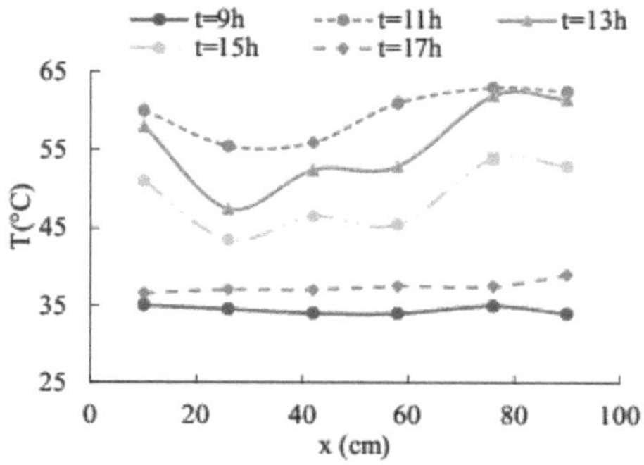

Figure 4.52. Variation of DS18b20 temperatures along the (ox) axis for the day on 02 July 2021

Figure 4.53 shows the evolution of the PV panel temperature for the day of 03 July 2021 carried out in PV collector mode with cooling every half hour. From these results, it can be seen that the temperatures of the six DS18b20 sensor positions show low intensities at 9 a.m. until they reach the minimum value T=32°C.

Temperatures then increased between 11 a.m. and 1 p.m., reaching a maximum value of T=65°C at 1 p.m. At 3pm, the temperature fell until it reached a minimum value of T=45°C at t=15h. At the end of the day, there was a continuous, non-acute drop in temperature until it reached a value of T=33°C at 5pm. Furthermore, there was no remarkable difference between the operation of the different positions of the DS18b20 sensors on the three days of the experiment.

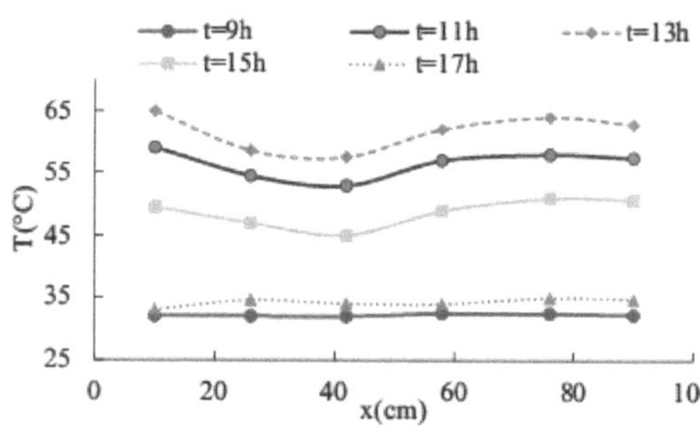

Figure 4.53. Variation in DS18b20 temperatures as a function of the (ox) axis for the day on 03 July 2021.

4.3.2 Temperatures along the (oz) axis

Figure 4.54 shows the evolution of the PV sensor temperatures in uncooled mode for the first day on 01 July 2021.From these results, it can be seen that the temperatures of the five positions of the DS18b20 sensors are almost constant. They then increase to reach an equal maximum value T=57.5°C at t=15h for the first direction and T=56°C at t=11h for the second direction. Temperatures also increased between 11am and 3pm. At 1pm, the temperature at fell until it reached a minimum value of T=54.5°C for the first direction and T=53°C for the second at t=13pm. At the end of the day, there was a continuous, non-acute decrease in temperature until it reached a value of T=37°C at 5pm.

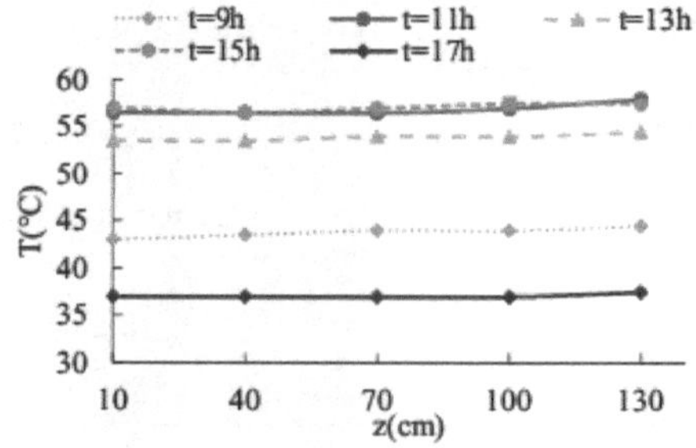

(a). First direction

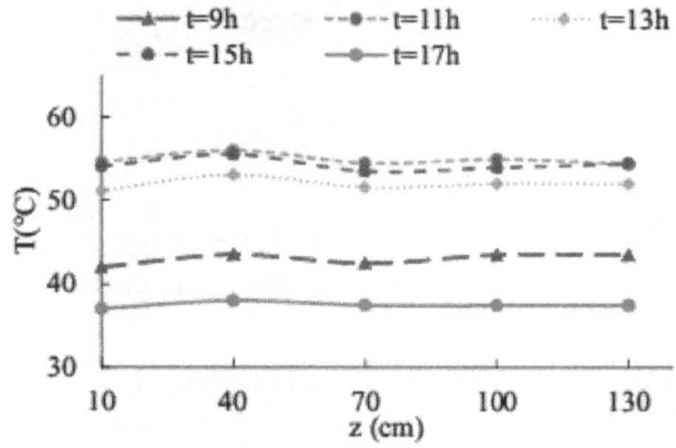

(b). Second direction

Figure 4.54. Temperature variation along the (oz) axis for 01 July 2021.

Figure 4.55 shows the evolution of the temperatures of the PV sensor in cooling mode every two hours for the second day of 02 July 2021. From these results, it can be seen that the temperatures of the five positions of the DS18b20 sensors are almost constant with low values at 9 a.m. until they reach the minimum value equal to T=32°C for the first direction and T=35°C for the second direction.

They then increase to reach a maximum value of T=63.5°C at t=13h for the first direction and T=59.5°C at t=11h for the second direction. Temperatures then increased between 11am and 1pm. At 3pm, the temperature fell until it reached a minimum value of T=45.5°C for the first direction and T=50.5°C for the second at t=15pm. At the end of the day, there was a continuous, non-acute decrease in temperature until reaching a value of T=37.5°C for the first direction and T=36.5°C for the second direction at 5pm.

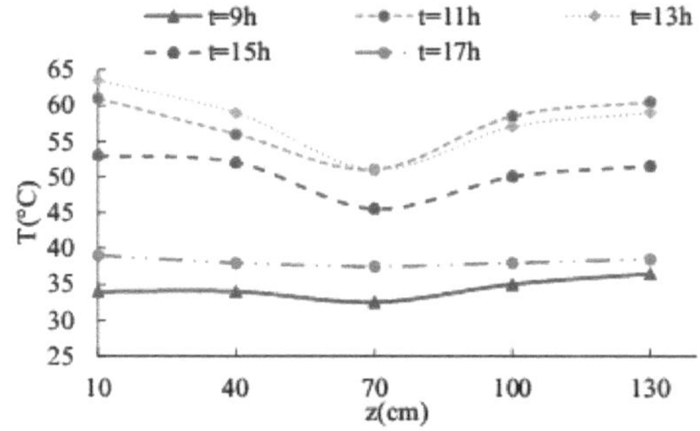

(a). première direction

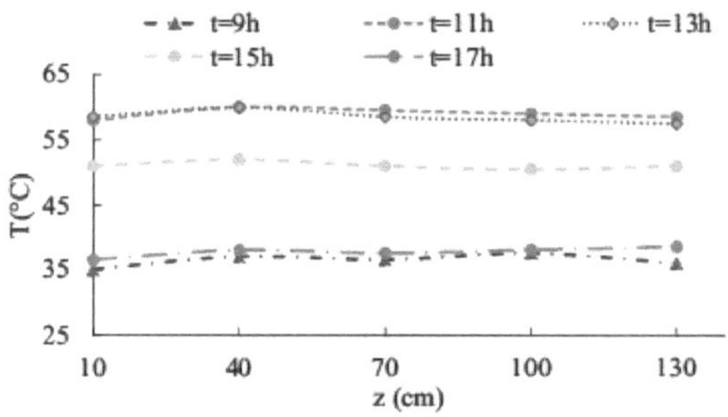

(b). deuxième direction

(b). second direction

Figure 4.55. Temperature variation along the axis (oz) for 02 July 2021.

Figure 4.56 shows the evolution of the temperatures of the PV sensor in cooling mode every half hour for the third day of 03 July 2021.From these results, it can be seen that the temperatures of the DS18b20 sensor positions are almost constant. We note that the presence of low values at 9 a.m. until reaching the minimum value is equal to T= 32 °C. Then they increase to reach a maximum value equal to T=64.5°C at t=13h for the first direction and T=66.5°C at t=13h for the second direction. Temperatures then increased between 11am and 3pm. At 3pm, the temperature fell until it reached a minimum value of T=45.5°C for the first direction and T=49°C for the second at t=15pm.

At the end of the day, there was a continuous, non-acute drop in temperature until it reached a value of T=33°C at 5pm. Furthermore, there is no remarkable difference between the operation of the different positions of the DS18b20 sensors.

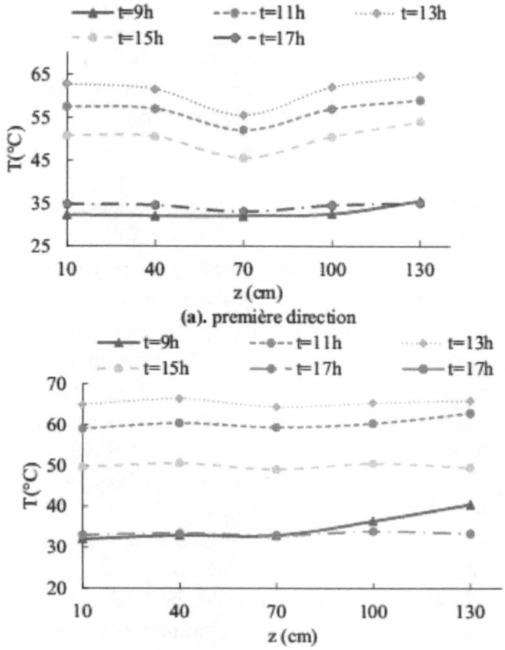

(b). deuxième direction

Figure 4.56. Variation of temperatures as a function of the axis (oz) for 03 July 2021.

5. Conclusion

In this chapter we have carried out a theoretical study on the photovoltaic panel and the closed chamber at the LAZEM laboratory at the National Engineering School, starting with an evaluation of the energy recovered by the panel over three days with a south-facing orientation and 30° tilt.

Secondly, we explained the steps involved in carrying out our experiment, and described the equipment needed to evaluate the power, efficiency and average temperature and sunshine generated by our panel in uncooled mode and with cooling every two hours and every half hour.

So, for the second experiment in a closed room for four days, the first in natural ventilation mode and the remaining days in air-conditioning mode at T=20°C, T=26°C and T=32°C, we presented the necessary equipment. We then followed the distribution of temperature and humidity and finally deduced a general equation for determining the point temperature as a function of the two parameters, the average temperature set on the air conditioner and the z point in m.

General conclusion

In just a few years, photovoltaic solar energy has come halfway from its modest beginnings to being used to generate electricity. It is an inexhaustible form of energy, since it is generated directly from the sun, and is an efficient, renewable form of energy available on today's market. It can also be used in buildings. Today, this technology enables grid-connected photovoltaic electricity production by integrating buildings and transforming roofs and facades into energy producers. These building-integrated photovoltaic systems are the only construction materials offering a return on investment. What's more, the diverse uses of PVIB systems offer architects and building designers plenty of scope for embellishing buildings. At the same time, building owners benefit from lower electricity bills and a green, innovative image. In this context, we are interested in the technical-economic study of the value of integrating photovoltaic systems into vertical buildings. More specifically, we referred to the building of the mechanical engineering department at the national engineering school in Sfax.

In the first part, we designed a grid-connected photovoltaic installation integrated on the roof of the mechanical engineering department. We also designed a photovoltaic installation in two different situations corresponding to the roof and facades of the LASEM laboratory. This led us to calculate the electrical requirements, enabling us to choose the photovoltaic components to be installed. The comparison between these two cases has shown that the installation of solar panels on the roof minimises the cost.

In the second part, we are interested in the development of a photovoltaic system and its installation at ENIS. This experiment enabled us to monitor its performance and heating under different operating conditions. The main focus is on the mode without and with cooling of the photovoltaic panel. The results showed that there is a great deal of waste compared with the energy gain. In the second part, the monitoring of parameters influencing the thermal comfort of a building using an acquisition system with an Arduino Uno card has already been developed in order to monitor changes in temperature and humidity in the room.

In the future, we propose to monitor the work involved in installing the photovoltaic system planned for the mechanical engineering department at the Sfax National Engineering School.

References

1. Christopher Gross July 2013 Project "Promotion of renewable energies and energy efficiency in Tunisia" Deutsche Gesellschaft für Internationale Zusammenarbeit (GIZ) GmbH
2. Stéphane Vighetti. Grid-connected photovoltaic systems: Selection and sizing of conversion stages. Sciences de l'ingénieur [physics]. Institut National Polytechnique de Grenoble - INPG, 2010. French. fftel-00525110
3. https://www.rncan.gc.ca/sites/www.nrcan.gc.ca/files/energy/pdf/solar-photovoltaic/NRCan BIPV Factsheet FR.pdf
4. https://www.steg.com.tn/dwl/prosol/Referentiel technique des installations PV.pdf

Kolibri 360 GmbH
Ferdinand-Jühlke-Straße 7
99095 Erfurt, Deutschland
produktsicherheit@kolibri360.de

Druck:
CPI Druckdienstleistungen GmbH
im Auftrag der
Kolibri 360 GmbH
Ferdinand-Jühlke-Str. 7
99095 Erfurt